FRAMEWORK FOR OUR FUTURE

Abridged Version of
Viability, Complexity and Us

FRAMEWORK FOR OUR FUTURE

Abridged Version of
Viability, Complexity and Us

JOHN KNIGHT

Cover design: J. Knight and M. Hite
Cover: Sunflower image after
https://commons.wikimedia.org/wiki/File:Sunflower_head.jpg,
https://creativecommons.org/licenses/by/3.0/deed.en

ISBN 978-1-77180-393-9 (paperback)
ISBN 978-1-77180-394-6 (epub)
ISBN 978-1-77180-395-3 (Kindle)

This is an original print edition of *Framework for Our Future: Abridged Version of* Viability, Complexity and Us.

TABLE OF CONTENTS

INTRODUCTION 1

CHAPTER 1   OUR WORLD 4

CHAPTER 2   OUR CURRENT CIRCUMSTANCE 19

CHAPTER 3   HOW WE GOT HERE 28

CHAPTER 4   WHAT THE FUTURE HOLDS 48

CHAPTER 5   AN ALTERNATIVE FUTURE 64

CHAPTER 6   ACHIEVING VIABILITY 79

CHAPTER 7   THE OUTLOOK 92

LIST OF FIGURES

Representations of the Human-Environment System

FIGURE 1   RESOURCE PERSPECTIVE 9

FIGURE 2   SYSTEM PERSPECTIVE 37

FIGURE 3   HUMAN PERSPECTIVE 65

Dedicated to those who are trying

INTRODUCTION

Today we have unprecedented amounts of information, technology and intellectual power at our fingertips. Yet we still argue incessantly about our current circumstance: its nature, its significance, its causes, the cures and the constraints. In doing so, we behave like the gentleman found wandering around under a streetlight late one night.

"Can I help?" a passerby asks.

"Yes," comes the reply, "I'm looking for my keys."

"Where did you last see them?"

"Somewhere out there," the gentleman says, waving vaguely beyond the circle of light.

"Then why are you looking here?"

"The light is better."

In order to respond appropriately to our current circumstance, we need to look for the keys where they are most likely to be found, and provide the light to do so. We require a framework that can help us both gain a comprehensive, coherent appreciation for our current circumstance and consider our possible futures. It should also help us evaluate possible solutions to issues, select the most appropriate and implement them. This book outlines such a framework.

To be useful, the framework must reflect who we are, the nature of our world, and our participation in it. It must also be easily incorporated into our decision making. Unfortunately, we find it difficult to relate to the earth's complexity and to come to terms with our nature, which hinders our efforts to create this framework.

For example, by contemplating a complex system, such as our world, from a number of perspectives and then merging them, we can more easily appreciate the system's structure and the complexity of how it functions as a whole. Indeed, doing so is a critical part of appreciating our current circumstance. Unfortunately, merging multiple perspectives is a challenge for us. Instead, we each tend to stake out a different preferred perspective of the complex system and adopt an incomplete and simplified view of how it functions. This contributes to our ongoing arguments about our current circumstance and to us searching for solutions where we would prefer to find them.

Similarly, it is possible to gain an appreciation for a complex system's meshwork of interconnections by reading a few descriptions of it, each from a different perspective. However, to make the connections between the perspectives clear, common features must be present in each description. This repetition detracts from the reading experience, and the reader is still left to complete the merging. In an effort to address these types of issues, this book takes the following approach.

The chapters are laid out in a logical order as pieces of a linear story, while still reflecting the many interconnected perspectives needed to tell it. "Think" statements are used as reminders to make those interconnections. Chapter 1 focuses on broadening our intellectual understanding of complex systems and our practical appreciation for real-world complex systems. Chapter 2 describes our current circumstance and introduces the idea of a complex human-environment system. In chapter 3, three factors of the complex human-environment system to which we can easily relate are used to outline a model of it. The model is then used to describe how we arrived at our current circumstance. In chapter 4, the model is used, in conjunction with the history of past complex cultures, to provide us with a window into our possible futures. Chapter 5 is devoted to describing an alternative future, one in which our culture will be functional over the long term (viable), and the challenges of reaching for it. Chapter 6 describes methods that can help each of us more easily make sense of ourselves and our world, apply the model and reach for viability.

When taken together, these chapters provide a framework, and a way to use it, that helps us to appreciate our circumstance, search for appropriate responses and implement them. Indeed, its application influenced the final chapter, which provides an outlook for the future.

Note that many of the claims made in this book are not fully justified, and there is insufficient detail to easily personalize and assimilate the framework, our current circumstance, the nature of viability and the changes needed to favour it. In particular, the book doesn't provide the examples and discussion needed for the reader to easily make the critical personal changes that are required if we are to strive toward viability.

That is not an oversight. This book is an abridged or summary version of *Viability, Complexity and Us: The Human-Environment System Constraining Our Future* (Knight 2020), which does provide the needed detail. Its introduction guides the reader to select those of its 18 stand-alone chapters that are relevant to their personal journey. It also provides a glossary of terms and a more comprehensive list of references.

Finally, although Western culture will be used to illustrate the points being made, the conclusions are not restricted to it. *Short* or *immediate term* refers to a period up to 15 years from 2015, an arbitrary reference year; *medium term,* up to 60 years (roughly one lifespan) from 2015; and *long term,* beyond 60 years from 2015.

Chapter 1

OUR WORLD

A useful framework will help us to more fully grasp our complex world's structure, how it functions, our place in it and how we affect its functioning. This means that it should, at its core, increase our familiarity with complexity and complex systems, both in the abstract and in real life. To that end, this chapter describes an abstract generic complex system and the real-world complex systems that will be referred to in later chapters: ecosystems, the climate system, the environment system (the biosphere) and us.

Complex Systems

The following description of a generic complex system may seem daunting, but don't despair. It will become clearer and more meaningful as you read the descriptions of real-world complex systems that immediately follow and the summary at the end of the chapter.

A complex system is characterized by its many variables and the wide variety of interactions among them. These linear and non-linear interactions take place over a broad range of time, space and intensity. As a result of these interactions, not only is a variable being routinely prompted to change, but it is prompting (influencing) the other variables it interacts with to change. Some of these interactions form feedback loops in which a change in a variable can, through a number of intermediate interactions, end up influencing the variable itself. Overall, at any given time, each of the system's variables is inevitably changing by some amount and speed in response to each of the various influences on it: a complex system is always in a state of flux.

That is how a complex system functions and changes, which is referred to as its dynamics. In the process, its structure is formed. The overall state and dynamics of the system can be described by the proportion of deterministic (simple cause-effect), random and chaotic features that it displays at a particular scale. But the details should not be forgotten. For example, the complexity of its dynamics means that it can display more features than expected from the simple sum of its parts, such as the ability to self-organize and "adapt."

We are helped to appreciate the details of a system's dynamics by using commonly understood terms to broadly characterize each of the influences on its variables, the variables' responses and the resulting overall change in the system's state: *drift, adjustment, preconditioned,*[1] *unexpected, directed, inertia, delay, imposed, complex/simple* and *external/internal.* For example, it takes time for a variable to respond to an influence on it. This can become a delay in the propagation of its response through the system. It is thus possible for one part of the system to be changing through a different process than another part, or through the same process but at a different rate and intensity. The delay can also be thought of as inertia to change.

The nature of a complex system's dynamics means that a particular change (event/effect) that the system displays is less likely to be the result of a single cause. It is more likely to be the product of many (direct plus indirect) causes that each occur at different scales of time, space and intensity, and differ in their significance. This, and the presence of feedback, can make it difficult to decide what is a cause and what an effect. Fortunately, we can instead treat a change in the system as its response to a flexible hierarchy of ongoing, system-wide (internally generated and externally sourced) influences on it: a process. If we do, then the age-old question of whether the

---

[1] Preconditioning occurs when a mix of material or mental events unevenly distributed in the past collectively makes a particular future change more probable but not certain. The events can favour that future change but not cause it.

chicken or the egg came first becomes meaningless. They were formed together, incrementally, over time, by a process.

In this context, changing is just a different view of functioning. Whether *change* refers to a normal part of the system's functioning or its deviation from that "normal" depends on whether an influence is internal or external in origin, and on the answer to the question, change relative to what? That is, what reference is being used to look at the system (e.g., how far in the past should we choose the reference state)? Failing to make these distinctions is a source of much of our arguing about the present and future.

Regardless of how we view its functioning/changing, if the system is to continue to exist, then its dynamics must direct it toward a stable state. But the system is always in a state of flux at some scale, so it rarely, if ever, reaches that state at all scales. A complex system's current state can therefore be described as partly out of equilibrium, constantly moving toward (catching up to) yesterday's theoretical equilibrium state or constantly catching up to ongoing changes: it exists in a dynamic, not fixed, state. A complex system exists in a quasi-stable state.

It will remain in that quasi-stable state for as long as its feedbacks favour that state. The degree to which a system can change and still remain in that state is limited by a collection of constraints, a somewhat flexible boundary around its current state. If the cumulative influences on the system result in it significantly crossing that boundary for long enough, then its feedbacks can instead favour it transitioning to a different near-equilibrium condition: an alternate quasi-stable state. The system will switch between quasi-stable states. Critically, the more the system is disrupted, the more difficult it is to predict what that future state will be. And, if the disruption results in a poorly restrained amplifying feedback in the system, then it can enter a highly unstable, even chaotic state.

With all of this in mind, it is clear that what we see in a complex system depends on the scale we use, the attention we pay to the features we are aware of and the context and reference we choose: that is, the perspective we take. There are many perspectives we can take of either the whole system or its parts. Each perspective can be valid, but on its own it is an incomplete view of the system. To gain a more complete view, we must coherently integrate those perspectives. By applying this discussion and its cautions when we contemplate the complex systems that form our world, we can more easily appreciate its current state, and thus our current circumstance.

Ecosystems

An ecosystem is any place on the earth that contains the living features (e.g., individuals of a species) and non-living features (e.g., nutrients, air, water, sunlight) that are needed to form a functional system of life. The individuals in such an area interact with one another and the inanimate world to form the structure and functioning of their host ecosystem. It, in turn, provides the niches and services its members require to complete their lifeway.[2] This completes a feedback. A closer inspection reveals that the feedback also constrains the individuals and the host ecosystem they form. The ecosystem's characteristics are constrained by the features of its member species and their interactions. In contrast, a species can only be a member if the ecosystem provides a niche for it.

This knowledge changes my view of the forest near my house. When walking in it, I no longer see a random mix of species or a chaotic arrangement of imperfect, autonomous individuals constantly engaged in a fierce competition, surrounded by wasted resources. I now see a space that is fully occupied by a kaleidoscope of trees, underbrush, bugs and animals, living and interacting in many ways, both above and below ground, over a wide range of space, time and intensity. And, collectively, they display various forms of order. For example, their feeding relationships form a somewhat flexible trophic web (also known as a food chain). As they go about their lives, they are all actively participating in the functioning of their supporting community in ways that contribute to its ongoing presence and changing.

---

[2] An individual's lifeway is its life cycle plus its experience of living: for example, its nurturing and training of offspring, its perception of its surroundings, its choices and how it interacts with others of its species.

I realize too that the individual organisms are neither cast in an inflexible mould nor free to do as they please. There is some flexibility in their lifeway, and they can adapt to the influences on them. But this flexibility is constrained by who they are (their species) and what they need, and thus by their supporting community. It is no wonder that the functioning of the forest results in a patchy and dynamic distribution of resources, species and individuals. This is not an imperfection to be fixed, nor is it a barrier to be overcome in order to reach a better or more perfect ecosystem. It is just a normal aspect of the ecosystem's current quasi-stable state and its functioning.

An ecosystem, such as our forest, is a complex, functional whole. From this perspective, it is clear that we cannot permanently alter one part of it (either by addition, subtraction or disruption) without, in some way, changing many other parts. Neither is it realistic to find simple causal (deterministic) laws that define how the system functions. It is more important that we understand and appreciate the relationships that describe how the system functions: its processes and the constraints on them. Particular attention should be paid to those relationships that define the flexible boundary within which its individuals must remain if they are to more easily complete their lifeway and ensure that their host community will be functional for future generations, and vice versa.

The Climate System

The flickering of sunlight on the wet, rustling leaves in our forest reminds me of the earth's climate system. Simplistically, this system functions by sunlight heating the earth's land and oceans and, indirectly, its atmosphere. Because the earth is a sphere, they all become hotter around the equator than around the poles. This heat difference is reduced by global-scale air and ocean currents that display a convoluted pattern because they are affected by the earth's spin and topography. The result is the variety of the earth's climates and their pattern of distribution around the globe.

A more complete view of the climate system's functioning and changing (its dynamics) is gained by including more of its features. For example, changes in the earth's ice sheets and sea ice can alter the albedo (reflectivity) of the earth's surface. This changes the amount of sunlight that ends up as heat in the atmosphere. The change in its temperature alters the nature and distribution of the earth's climates (climate change), which, in turn, affects some of the climate system's features, such as its ice cover, thus completing a feedback.

Consider too that the oceans play a significant role in the climate system. Their large capacity to store heat and carbon dioxide (CO_2), and the transfer of both to and from the atmosphere, helps to stabilize the climate system (figure 1).

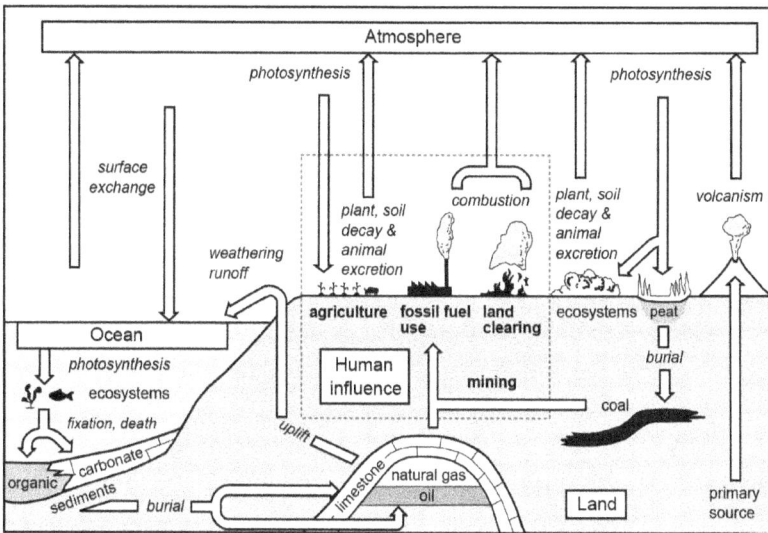

Figure 1. Simplified carbon cycle: a resource perspective of the human–environment system. The cycle is divided into atmosphere, ocean and land aspects. The dashed box outlines the human influence on the cycles. Italics represent processes and pathways. Non-italics outside the human influence box represent reservoirs; inside they represent activities. There are many forms of carbon. Where the reservoir is oxygen rich (e.g., the atmosphere), the dominant form is carbon dioxide. Where oxygen is absent (e.g., peat swamps and underground), then hydrocarbons, such as methane and oil, can exist. After Strahler, A. and Strahler, A. 1997. Physical geography: Science and systems of the human environment. John Wiley and Sons, New York. Figure 20.12.

But that transfer occurs slowly. Thus, if the change in the CO2 content and temperature of the atmosphere occurs faster than the oceans can effectively respond, then the earth's climates will change more quickly than expected.

Finally, the deviations in the earth's orbit around the sun from a perfect circle and the varying tilt of the earth's axis result in a very slow, ongoing change in the amount of sunlight reaching the earth (the Milankovitch sunlight cycle). This is responsible for the climate system's ~100,000-year cycle of switching between a quasi-stable glacial and an interglacial state over the last two million years.

The climate system is a dynamic complex system that exists in a quasi-stable state and displays deterministic, chaotic and random features. A particular aspect of the system's dynamics can be described by generic descriptors such as *inertia* (e.g., the resistance of the oceans to change), *delay*s (e.g., the time between a change in the atmosphere's temperature and changes in the ice sheets' volume) and *preconditioning*.

Thus, whether we think that a change in the climate system will result in significant changes to the earth's climates or not depends on which parts/variables of the system we consider, the references we use and the scale of time, space and intensity we choose. The following rule of thumb might help: the effect of a small but long-term change can have a long-term outcome that is just as significant as a very large but short-term change. But the details for each will be different.

The Environment System

The functioning of the earth's climate system influences the global distribution of sunlight, heat and available fresh water. These features all play key roles in the functioning of plants, which form the base of an ecosystem's trophic web. This is one of the many ways in which the climate system and ecosystems are linked together.

As a region's climate changes, the members of its ecosystems respond through a mix of migration (to follow their climate-influenced ecosystem niche) and adjusting their lifeway characteristics (to better match their new living conditions). Their responses change their host ecosystem's location, mix of species and functioning. Collectively, these changes alter the region's albedo and the amount of water added to and CO_2 removed from the atmosphere by the plants, which affects the region's climate. This climate-ecosystem feedback is just one of the feedbacks joining the earth's climate system and ecosystems together. The resource cycle for carbon illustrates some of them (figure 1).

These links and feedbacks inseparably bind the two systems together into a dynamic, complex environment system. Its functioning ensures that a change, regardless of scale, to one part will eventually spread to the other parts. Fortunately, the nature of the system's functioning allows it to maintain the system in a dynamic, quasi-stable state. However, if the influences on the system or its response to them exceed some threshold, then the environment will switch into a different quasi-stable state.

Indeed, as previously mentioned, over the last two million years, the Milankovitch sunlight cycle drove the earth's environment to switch back and forth between a glacial and an interglacial quasi-stable state some 20 times. However, the details of those switches, such as the speed and intensity of the changes, were mediated by the changes in the earth's snow and ice cover (albedo), its biomes (albedo and CO_2) and its oceans (dissolved CO_2 and heat). They, collectively, formed a feedback that amplified the initial consequences of the change in sunlight to create the large swings between the two states.

The environment system is also influenced by random events. They can be internally generated by the system's functioning or imposed from outside. Examples are radiation-induced genetic mutations, volcanic eruptions and asteroid strikes. We complete our lifeway as part of this environment system. But what is our role?

Our Place

The human species is certainly unique, but not in a way that separates us from the earth's environment system or its other species. After all, individual humans, like all organisms, only exist because our parts form a complex functional whole. Satisfying this requirement constrains our physical and lifeway features and, because we are social animals, our culture. Consider that the size of our organs (e.g., our heart) and many of our life-cycle characteristics (e.g., how long we sleep) correlate with our body mass (our size): our features are allometrically constrained. Similarly, we individual humans, like the members of all species, depend on our surroundings to complete our lifeway. Our need to use the environment also constrains our physical, lifeway and cultural features (figure 1). Humans are, like all other species, active participants in the earth's environment system.

Culture is important because it is a mechanism used by social species to help its members form and maintain stable and adaptable groups. As the members of a greater human group go about their day-to-day business, they interact with one another. This unavoidably results in them dividing themselves into overlapping, flexible subgroups (e.g., genders, trades, political parties and sports groups), in which they then participate. The interactions among the members of a subgroup are its functioning, which forms its structure, beliefs, behaviours and values: its culture. Similarly, the (formal and informal) interactions among a greater group's subgroups (in reality, among their respective members) establish the relationships and boundaries among those subgroups. This is the foundation of the greater group's culture.

The culture of a group is represented by its world view. It provides guidance to the culture's members about how and why to live in the world in a particular way. At its most basic, a culture's world view prompts its members to consider their individual and collective interests, how humans function, their supporting environment (their culture's environmental endowment) and the constraints on their lives arising from these factors.

A culture can guide its members because they embed its world view into their personal world view as they grow up: they are acculturated to accept its guidance. As adults, the members are largely unaware of the degree to which their culture's world view has become a part of their personal identity. For example, Westerners see seven colours in a rainbow; other cultures don't. Consider too that an adult member's response to and use of the cultural symbols they learned as a child is largely unconscious.

This mention of symbols is a reminder that a culture's visible features (such as its customs, dress and buildings) have both a practical and a symbolic purpose. When its features act as symbols, they remind the members of those aspects of their culture's world view that are hidden from view, such as its beliefs and values (the culture's norms). By the same token, the members use symbols and symbolic acts to represent their acceptance/recognition/rejection of these norms. Think of how a salute or bow shows deference, while a raised fist or finger indicates rejection. Thus symbols and symbolic acts consciously and unconsciously influence an individual member's views and decisions about themselves, their social identity, reality and normality. And they help the members to express those views. Therefore, symbols and symbolic acts both help to bind the members together as a culture and give it flexibility. They too are a part of a culture's functioning.

The members can change their culture's world view because they are both its carriers and its decision makers. So, they can adjust or maintain its guidance as they see fit; this is one reason why a culture can guide but not order its members. From this perspective, the functioning of a culture is its members' decision making, and their arguing about past decisions. But their decision making is influenced by their culture's world view and the symbolic representations of it. A culture's functioning thus includes a member-decision–cultural-world-view feedback loop (member–culture feedback).

This feedback combines with the culture's historical roots to endow the culture with inertia to change. This gives the culture stability, but it also constrains the members' ability to change it. Consequently, although a culture's current generation, like its past ones, does adjust their culture's world view, its core only changes slowly, on a timescale of at least a few generations. There are limits to how fast a culture can change and remain functional.

It is easy to conclude that a culture and its members form a complex cultural system that exists in a dynamic, quasi-stable state. We thus participate in the environment system both as individuals and as members of a culture: as part of a cultural system.

The Diversity of Cultures

There is one other feature of human cultures that plays a key role in the framework: their diversity. Each culture is unique; however, those cultures whose population falls within a particular size range are seen to have similar basic features, such as their organizational structure and the essence of their world view. As a result, the diversity of human cultures can be divided into types.

In broad terms, hunter-gatherer cultures have small populations (bands), display an egalitarian power structure and use simple processes to secure resources, reaffirm their world view and ensure unity. Their organizational structure is simple: a simple culture.[3] (The term *simple culture* will be used throughout to refer to a culture whose organizational structural is simple.) In contrast, state cultures have large populations (some of whom live in urban centres), are dependent on agriculture and can be industrialized. These cultures display a hierarchical power structure that, simplistically, divides the members into public and elite.[4] To maintain unity, they engage in a complex process that reaffirms their culture's world view and authority structure, and ensures that the structure remains functional. Their reaffirmation and authority-maintenance efforts include ceremonies and features such as voting, paying taxes (as services, goods or cash) and receiving benefits (as services, goods or cash). The organizational

[3] All cultures are complex systems. The terms *simple culture* and *complex culture* only refer to the nature of the culture's organizational structure: its level of structural complexity.

[4] The elite are those with significantly more power, wealth or influence than average. For example, they are our social, political, economic, moral, military, religious, educational, business and technological leaders. Because of this, they are burdened with more responsibilities than the public to ensure that their culture remains functional over the long term.

structure of a state culture is complex: a complex culture.[5] (The term *complex culture* will be used throughout to refer to a culture whose organizational structure is complex.)

The characteristics of the more complex of the hunter-gatherer and the less complex of the state-agricultural cultures lie in between. They display an intermediate ranked structure and use a reaffirmation process that is intermediate in complexity.

Also varying with a culture's group size/structural complexity is the essence of its world view: its guidance to its members about the nature of the world, how to live in it and why. The members of simple cultures tend to see the world as having a community-like structure and their relationship with their environment as largely one of being participants in a wider spiritual and material community. Simple cultures favour a low-resource lifestyle and a simple reaffirmation process, and thus have a low resource demand per capita. In contrast, the members of more complex cultures (state cultures) prefer to see the world as having a hierarchical structure and their relationship with their environment as one of being transformers and managers. Complex cultures favour a resource-intensive lifestyle and a complex reaffirmation process, and thus have a high resource demand per capita. The characteristics of the cultures on the cusp of becoming complex lie in between.

If we take a wider, longer-term view of the cultural types, then their characteristics can be arranged on a continuum. For example, work specialization ranges from very little, if any, in hunter-gatherer cultures, to some in settled hunter-gatherer or ranked farming cultures, to formalized specialization in complex farming cultures (city-states), to formalized and highly specialized in industrial cultures. In this view, cultures are seen to be related to one another. However, if we take a closer, shorter-term view of the cultures within a particular cultural type, then the differences between them are more marked, and cultures can appear to be independent of one another.

[5] See note 3 above.

A greater appreciation for the similarity and diversity of cultures and the connections among them appears when the wide and the close-up views are treated as different perspectives of a dynamic, complex cultural system. In this view, a culture is a functional system in which the interactions (including feedbacks) among its individual members, their culture's world view and their environmental endowment form and maintain the basic structure and functioning of their culture, and the core of its associated world view. Those interactions, and the influences on them, operate over a wide range of time, space and intensity, for example, from the time scale of the members' day-to-day lives to multiple generations.

The complex system view of culture also helps us appreciate how a culture changes both within and between types. In broad view, a culture's functioning allows its members to (consciously or unconsciously) make day-to-day decisions that can (directly or indirectly) change their culture. This can be just a peripheral change, such as the preferred clothing, or a major change, such as a change in the preferred god(s).

Some of these changes can, collectively, result in an increase in the culture's (structural) complexity. Such an increase is more likely if the members also change their culture's world view so that it guides them to routinely resolve issues in ways that favour an increase in complexity. Indeed, it is even possible for a self-sustaining complexity-favouring feedback loop to form. But the process of increasing a culture's complexity (i.e., changing between cultural types) is more convoluted than it sounds. There is inertia to making the needed changes. And the interconnected constraints on the culture (its functional boundary) limit the changes that can be made if the culture is to function as a whole.

Consider the broad changes the members of a hunter-gatherer culture must make to become a state culture dependent on farming. For example, they must adopt a complex culture's obligatory reaffirmation and authority maintenance efforts because that is how a complex culture maintains its functioning and the cohesion of its population. The cost per capita of this process is high compared to the equivalent process in a simple culture. Thus a hunter-gatherer culture can only become a functional complex culture if it can also pay for the extra fixed cost. This means having access to the natural resources that can be, ultimately, used to pay that cost. If the hunter-gatherer culture has that access, then it has another condition to meet. The members would have to change their relationship with the environment to favour the extraction of those resources and accept being part of a hierarchy: they must change their world view. The process of changing a culture from one type to another affects all aspects of the culture and takes time. It is best treated as a transformation from one quasi-stable cultural state to another.

*∗∗

In summary, chapter 1 first provided a generic description of complexity and complex systems. In particular, a complex system's many variables interact with one another, including feedback, over a wide range of time, space and intensity. This, its functioning, ensures that the system is always changing at some scale (it is quasi-stable) and that an event can have multiple causes. Overall, a complex system displays a mix of deterministic, random and chaotic features that is most easily understood by focusing on properties, processes, limits and constraints.

The world around us can be divided, for convenience, into complex systems such as the climate system and ecosystems. They interact with one another, so they also collectively form the complex global environment system, or biosphere. The human species, like all species, is a participant in the biosphere.

Humans are social animals who form groups. The members of a group interact to create a culture whose world view guides the members' decision making. A world view is constrained by human features such as how we make decisions and the need for the members to get along with one another. It is also constrained by environmental features, such as our reliance on them to complete our lifeway and maintain our culture.

Hunter-gatherer cultures are the least structurally complex of human cultures. Their members can, over generations, increase the structural complexity of their culture. But if they do, then they have to adjust their culture's world view to reflect the human and environmental constraints on their culture being functional. This is achieved by the members adjusting their relationship with one another (they form a power hierarchy) and with their environmental endowment (they come to favour transforming it). In essence, they transform their culture into a state culture.

However, a hierarchy comes with social complications, and our perception of the environment is biased by our human foibles, such as how we choose to look at it. Invariably, the result of becoming a more complex culture burdens the members with additional social and environmental responsibilities.

This context is key to understanding our current circumstance, how we arrived here, what the future might hold and how best to respond. All are subjects of the remaining chapters.

Chapter 2

OUR CURRENT CIRCUMSTANCE

The functioning of the environment system and our cultural systems, and our personal participation in both, have resulted in our current circumstance. This chapter discusses it from two perspectives: the biosphere's state and our resource wants.

The Biosphere's Current State

As we secure the resources we want and live our preferred lifestyle, we contribute to the state of the earth's biosphere. It is convenient to divide our contributions into three overlapping types: our direct actions (e.g., land alterations, exotic introductions, water diversion), pollution and climate warming. Their cumulative influence on the biosphere's functioning has resulted in the current state of its atmospheric, oceanic and land-based environments.

Climate warming is our most significant *indirect* contribution to the state of the biosphere (but not our only significant contribution). Consider that some of our pollution (e.g., greenhouse gases) and direct actions (e.g., land clearing) affect the earth's climate system. They change the earth's albedo and the composition of the atmosphere in ways that result in the warming of the lower atmosphere. But the products of climate warming (e.g., the melting of sea ice, ice sheets and permafrost) can also change the earth's albedo and the composition of the atmosphere in ways that warm the lower atmosphere. Feedbacks are part of our contribution to the current state of the biosphere.

As the atmosphere warms, the earth's climate zones and the oceans' biozones change and shift poleward, which individuals and thus species respond to by migrating and changing their lifeway characteristics. But their efforts to do so are affected by other aspects of our direct actions and pollution. For example, atmospheric pollution is transferred to the oceans, which changes the environment in which marine organisms live (e.g., by making it more acidic). These changes affect ocean ecosystems at the same time as their trophic webs are being affected from the bottom up by our marine pollution from the land (e.g., plastic and soluble waste chemicals), and from the top down by our direct actions (e.g., fishing). Equivalent processes are affecting land ecosystems.

The overall result of our contributions is the global-scale disruption of the biosphere's functioning. To personalize this disruption, consider the degree to which the earth's surface and the ecosystems it hosts have been disrupted or destroyed by our direct actions.

By 2000, humans had appropriated about 34% of the earth's ice-free land surface for agriculture (excluding tree plantations). One-third (12%) was used for crop land, and two-thirds (22%) provided pasture for our domesticated animals. To create our crop land, we largely destroyed the pre-existing ecosystems. They include 30% of the pre-farming temperate forest biome in Europe and the United States; significant portions of tropical deciduous forests in South Asia; and a significant proportion of global grasslands. To create just our pastures, we significantly disrupted some 50% of the pre-farming grassland biome and substantial portions of the savannah and shrubland biomes.[6] Today, the last remnants of the ecosystems hosting 44% of the world's terrestrial plant species and 35% of its terrestrial vertebrate species are found in 25 biodiversity hotspots. These areas make up only 1.4% of the earth's surface.[7]

[6] Ramankutty, N., Evan, A., Monfreda, C. and Foley, J. A. 2008. Farming the planet: 1. Geographic distribution of global agricultural lands in the year 2000, Global Biogeochemical Cycles, vol. 22, GB1003.

[7] Myers, N. 2002. A convincing call for conservation. Review of The future of life, by Wilson, E. O. Science, vol. 295, pp. 447–448.

There is every indication that marine ecosystems have also been significantly disrupted and destroyed on a global scale. For example, by 2008, at least 84% of industrial fisheries (coastal and open ocean) were fully exploited, overexploited or collapsed, implying that their host ecosystems have been significantly disrupted.[8]

The above percentages do not fully capture the degree to which we have disrupted and destroyed the biosphere's ecosystems because they only consider the most obvious of our direct contributions. The percentages do not include our pollution or our climate warming contributions. Nor should we forget that these changes are trade-offs in a finite system, not isolated impacts, and they have collateral consequences.

The above changes to the biosphere could be seen as a normal part of environmental change. However, the intensity, scale and speed at which they are taking place, especially over the last 200 years, marks them as unusual, even on a geological time scale. So does the fact that a single species, us, is largely responsible for the changes.

Regardless of the perspective we take, the current state of the earth's biosphere is highly anomalous compared to just 200 years ago, let alone to 12,000 years ago, before we took up farming. We have disrupted it to such a degree that there is no longer a distinction between pristine wilderness and human-affected areas. We have transformed the biosphere into a human-dominated environment system.

However, we experience difficulty taking this reality into account. This is illustrated by one study that tried to comprehensively describe the biosphere's current state. In it, the biosphere's functioning was represented by nine factors: biodiversity loss, climate warming, the nitrogen cycle, stratospheric ozone depletion, ocean acidification, global freshwater use, change in land use, chemical pollution and atmospheric loading with aerosols. A threshold for each factor was set, which, if crossed, was thought to indicate that the biosphere's

---

[8] Froese, R., Zeller, D., Kleisner, K. and Pauly, D. 2012. What catch data can tell us about the status of global fisheries. Marine Biology, vol. 159, no. 6, pp. 1283–1292.

functioning is being critically compromised (severely disrupted). The authors suggested that our activities have already driven the factors of biodiversity loss, climate warming and the nitrogen cycle across their threshold. The status of each of the remaining factors is either "yet to be determined," "satisfactory" or rising toward its threshold.

The idea of thresholds is worthy of our attention because it accepts that the environment exists in a quasi-stable state and that there are environmental constraints on how we can live. It also accepts that our changes to the biosphere are significant. Indeed, the current status of these factors informs us of our lack of awareness of their constraints and our lack of attention to their importance. However, the study's conclusions suffer from significant limitations.

One limitation is the authors' confidence that the dynamics of the complex biosphere's functioning and its state can be meaningfully or satisfactorily represented by nine numerical factors each with a largely fixed threshold. Another is that these factors are mostly considered in isolation. Perhaps the most significant limitation is the conclusion: "The evidence so far suggests that, as long as the thresholds are not crossed, humanity has the freedom to pursue long-term social and economic development."[9] It implies that our current lifestyle and goals can be maintained, if we become better biosphere managers. Why this is unrealistic is a thread that passes through the following chapters.

Satisfying Our Current Resource Wants

A casual consideration of our circumstance allows us to conclude that the earth is perfectly able to satisfy our current natural resource demands. In particular, the earth is currently most able to supply us with metals and minerals, and farmed food. It is least able to supply us with wild foods. Certainly, there are some regional-scale supply difficulties, particularly for fresh water and health services, but there

[9] Rockström, J., Steffen, W., Noone, K., Persson, Å., Chapin, F. S. et al. 2009. A safe operating space for humanity. Nature, vol. 461, pp. 472–475.

are currently no acute, and only relatively minor chronic, global-scale resource shortages. Any problems that do arise appear to be the result of human-caused inefficiencies and inequalities: poor management.

However, on closer inspection, a different picture emerges. Consider that energy is needed to supply resources (including energy). Thus the amount of energy available for us to supply other resources is constrained by the energy its takes to supply energy: the ratio of energy return on energy invested (EROI) to make energy available must be above one. Consider too that supplying a resource also uses resources other than energy. Thus, for example, by increasing the supply of one resource, such as energy, by using, say, biofuels from corn, we affect the supply of other resources, such as agricultural land and water, and thus food. The processes of supplying resources are interconnected.

The biosphere's functioning, directly and indirectly, ultimately provides all of the resources (including services) we want. However, the consequences of us supplying, using and discarding resources can disrupt its functioning. Once these non-economic costs are sufficiently high, they interfere with the biosphere's ability to provide the resources we want, where we expect them. Our efforts to maintain its functioning require resources, which completes a feedback. On a global scale, importing a resource doesn't eliminate these non-economic costs because the exporter still absorbs most of them, while transporting the resource adds to the costs.

We would therefore gain a more realistic view of satisfying our resource wants if we were to conduct a more comprehensive assessment of their availability, one equivalent to a lifecycle analysis. It would include a more complete range of the factors involved: for example, the full range of the resources needed to secure a resource; the non-economic costs of supplying, using and disposing of resources; our ability to maintain the biosphere's functioning and the resource costs of doing so; and the role of cultural beliefs. It would also include the connections/interactions among the factors, especially feedbacks.

Unfortunately, we find it difficult to conceptualize and relate to the complexity of our resource circumstance in such a comprehensive manner. We preferentially deal with this difficulty by making simplifying assumptions about our circumstance. The footprint model illustrates the benefits and pitfalls of doing so. It consolidates the environmental functions needed to both supply the resources to a human group (town, city, country, etc.) and dispose of all of its waste into a single hypothetical area of land called a footprint. The difference between the size of a group's footprint and the actual amount of land available to its members is an estimate of whether or not the group is satisfying their resource-supply and waste-disposal needs in a sustainable manner.[10] For example, the model unsurprisingly indicates that the footprint of cities is so much larger than the physical area they occupy that, for their cities to function, urbanites have to be net importers of resources and net exporters of waste.[11] And a comparison of footprint sizes (i.e., the relative difference in their sizes) illustrates the relative changes in the sustainability of our wants. For example, the size of our global footprint has increased from 70% of the biosphere's estimated ability to supply our wants in 1961 to 120% in 1999.[12] This indicates that our demand for resources and services has undergone rapid, significant growth. It also, less meaningfully, suggests that by 1999 we had exceeded the earth's carrying capacity for humans.

"Less meaningfully" is emphasized because there are significant limitations to implementing the footprint model. Consider that it combines things that should be left separate, such as sustainable and unsustainable land use; it concentrates on political regions not bioregions to calculate footprints; and it is unable to measure or include

[10] Rees, D. A. 1993. Time for scientists to pay their dues. Nature, vol. 363, pp. 203–204.

[11] Wackernagel, M. and Rees, W. 1996. Our ecological footprint. New Society Publishers, Gabriola.

[12] Wackernagel, M., Schulz, N. B., Deumling, D., Linares, A. C., Jenkins, M. et al. 2002. Tracking the ecological overshoot of the human economy. Proceedings of the National Academy of Sciences, vol. 99, no. 14, pp. 9266–9271.

planning decisions. Its aggregations also mask the interconnections between factors and the importance of the time frame used.[13]

Despite the limitations, the general conclusion that our resource demand is increasing and that there are limits to their supply cannot be dismissed as a fringe fantasy from a flaky model because it is supported by independent evidence. For example, allometric relationships indicate that, for our body size, there are between 80 and 800 times more humans living today than the environment could support when all humans lived as hunter-gatherers, which was our way of life for some 200,000 years. In order to support our current population and its per person demand for resources, which are both much larger than those of our hunter-gatherer ancestors, we have had to, from the environment's perspective, significantly disrupt its functioning. We likely have exceeded, or are in the process of exceeding (overshooting), the earth's longer-term carrying capacity for humans.

Although making simplifying assumptions can provide us with broad insight into the availability of resources, it can also hinder our efforts to comprehensively assess that availability. For example, simplification favours our tendency to treat a limited selection or an aggregate of variables as if they fully represent a resource's availability, or the state of our biosphere's functioning that provides it. Think of a resource footprint, the practice of economics or a threshold for one of the biosphere's features, such as the atmosphere's average temperature. This focus tends to distract us from the complexity and dynamics of supply, demand and the biosphere's functioning. Thus, without care, making simplifying assumptions can hinder our efforts to think about our current resource circumstance in a more comprehensive and coherent manner, and do so at a personal level.

If we do devote the time, effort and care, then we can at least come to personally appreciate that the connections and interactions between the factors affecting resource availability strongly constrain the amount of resources that we can routinely secure over the long

[13] Jeroen, C. J. M., Van den Bergh, J. C. M. and Verbruggen, H. 1999. Spatial sustainability, trade and indicators: An evaluation of the 'ecological footprint'. Ecological Economics, vol. 29, no. 1, pp. 61–72.

term. In particular, we can accept that we don't secure our resource wants or dispose of our waste by tapping into excess or unused environmental resources (including services). They simply don't exist. Indeed, the ability of the environment's functioning to provide resources is constrained, and thus most of the earth's resources we want are finite (in a practical sense). Think of the limits to renewable resources. This brings with it the realization that, in essence, our resource circumstance is the product of our limited recognition of the impacts, compromises and trade-offs we inevitably make in order to secure them.

For example, from the biosphere's perspective, when our ancestors took up farming some 12,000 years ago, they traded its self-sustaining, human-supporting indigenous ecosystems for agricultural ecosystems. These agricultural ecosystems consisted of our ancestors plus the plants and animals that fed them. The functioning of agricultural ecosystems depends on humans, so we became a keystone species. The price the biosphere paid for this trade-off included more than the disruption and destruction of the indigenous ecosystems needed to create our fields and pastures. Its functioning was also changed by the collateral effects resulting from the presence of agricultural ecosystems, the increased presence of our ancestors and their greater involvement in biosphere matters. Think pest control.

From our ancestors' perspective, by taking up farming, they relieved themselves of the vagaries of a limited seasonal food supply. In exchange, they became responsible for the lives of their food plants and animals, and for the fields on which they grow. And, whether they knew it or not, our ancestors became responsible for their (direct and collateral) disruptions of the biosphere's functioning. They also had to change their culture's world view so that it supported their transformation of the land and prompted them to discharge their expanded responsibilities to their agricultural ecosystems, their environmental endowment and to one another. This was the price (responsibilities, trade-offs, compromises and more) that humans paid when they adopted agriculture to satisfy their food wants.

These types of changes and trade-offs lie behind our current efforts to satisfy today's resource demands. When contemplating our current resource circumstance, we need to consider more than our ability to secure them. We need to include factors such as our (direct and indirect) impacts on the biosphere's functioning from extracting, using and discarding them; the responsibilities we gain; and our beliefs and feedbacks.

Yes, it is true that, with the application of technology, we have maintained our desired supply of resources. However, the cumulative impacts from our extracting, using and discarding them have, by any measure, reduced the biosphere's ability to provide our resource wants, and dispose of our waste, in the amounts and at the speeds we demand. The biosphere is now a human-dominated system and the availability of resources from it depends on how we discharge our obligations to it.

***

In summary, chapter 2 revealed that our successful efforts to satisfy our resource wants have resulted in our environmental endowment being disrupted to such a degree that the biosphere is now a human-dominated system. In the process, we have become increasingly responsible for ensuring that its functioning can provide us with our resource wants. By default, we have become environmental and human managers.

Chapter 3

HOW WE GOT HERE

It is easy to blame our current circumstance on an increase in either our population or our demand for resources, or both. However, if we think of our circumstance as being part of a complex human-environment system, then we require a more nuanced and comprehensive explanation: one that includes our individual and cultural characteristics (our "foibles"), the environment and complexity. This chapter addresses that need by first outlining the three primary factors of the complex human-environment system and how they have contributed to our circumstance. It then integrates them to create a comprehensive descriptive model of the complex human-environment system. Applying the model provides us with the required explanation.

The Decision Makers: Individuals

Our awareness of the world and our decision making about it is a product of our brain and its host body: a complex system.

Our brain's functioning creates a reality for us (our internal reality) by using a mix of the information in our memory (including feelings and our virtual body image) and our sensing of both the state of our body and the world around us. Most of our internal reality and the processes that create it are unavailable to us directly: they reside in our unconscious. Fortunately, extensions of the unconscious allow us to consciously experience a small part of both. Although small, it is enough for us to knowingly live our life and make conscious decisions.

However, both the nature of our brain's functioning and the influences on it ensure that our internal reality is an imperfect and incomplete representation of ourselves and our surroundings. It imperfectly represents the reality that exists beyond itself: the external reality (the one that remains after we die). The differences between our internal reality and the external reality can be thought of as our internal reality's divergences. Our brain doesn't automatically make us aware of these divergences.

More broadly, although our conscious experiences and decision making are wonderfully real and coherent to us, they still reflect our brain's limitations. Fortunately, under many circumstances, this has no significant effect on our living in the world. However, because we exist at the boundary between the two realities, there are some circumstances where our brain's limitations can result in us experiencing serious unpleasant and unwanted outcomes. The brain's limitations thus add a burden to our consciousness: we are responsible for deciding which parts of our conscious awareness and thinking are true in the external reality, which are not and when the deviations/differences really matter. To appropriately discharge this responsibility, we are obliged to check what we think we know and our resulting decisions. This is a challenge for us because we innately tend to believe that our decisions are made in a logical and rational manner, and that the internal reality on which they are based is an accurate representation of the external reality.

To address this challenge, we must consciously take into account the nature of our brain's functioning, especially its (our) decision-making foibles. Our decision making occurs, mostly, in our unconscious, using processes that are automatic, fast and inescapable. An extension of those processes is our conscious (cognitive) decision-making processes, which occur more slowly and can be directed by us (e.g., by our use of logic). Although only a small part of our decision making occurs consciously, it does allow us to reconsider our unconscious decisions, when they appear in our conscious awareness. However, if our unconscious makes a firm decision, then it is usually accepted by our conscious. Only when an unconscious decision is uncertain, or we choose to question it, is it used as a starting point for our conscious decision making.

Unsurprisingly, the characteristics of our brain's conscious decision-making processes have much in common with the unconscious processes. For example, when we reason about information, we compare it to a reference. That reference can be a part of our normal suite of references, which is an aggregate of our innate and learned references, or created on the fly from memory as needed.

Our reasoning uses heuristic processes (rules of thumb) that treat information as more than fact. They also consider the amount of information about a topic recalled from memory, the ease/speed of recalling it and the emotion/feeling attached to it. These features collectively mean that our decision making can be influenced by many factors beyond what we might imagine, such as our personal experiences, context, our (conscious and innate) desires, the limitations of our senses, feelings, our membership in subgroups and our culture's world view.

Our decision-making foibles, the influences on them and the presence of divergences are all included in our (unconscious and conscious) decisions. Thus, although our personal decisions can be treated, from our perspective, as rational and logical, we should also accept that they are accompanied by significant limitations: uncertainty, as well as biases, mistakes and omissions (errors, if you like). Fortunately, we do have the capacity to recognize and take them into account, if we wish. We are therefore neither the emotional idiots targeted by marketers and advertisers, nor the rational, informed consumers so beloved by economic theorists. We are people with constrained abilities, faced with making decisions in a complex world.

Our decision-making foibles can be more fully appreciated and more easily accommodated if they are viewed within the context in which they developed. Our current mental abilities arose over hundreds of thousands of years. During this period, our ancestors formed small hunter-gatherer groups within which each person had close ties to all of the other members of the group. They lived close to the land, immersed in their supporting environment, with only limited material power to manipulate it. Thus, the diversity of topics about which they needed to make decisions, and the contexts in which they had to make them, were much fewer than today. It is only in the last six or so thousand years (some 240 generations, 120 lifespans) that those topics and their contexts have come to include large populations in urban centres, complex cultures with a hierarchy, many more responsibilities and large amounts of information. It is only relatively recently that the technology that can significantly alter our lives and extensively modify the environment has become a ubiquitous presence in our lives. And today we have to make our decisions while under the pressure of inescapable, sophisticated, but faceless, human influences. Think of advertising and social media. Today we are applying our decision-making abilities in unfamiliar territory.

With this in mind, an individual's personal contribution to our current circumstance is illustrated by our relationship with complex systems, and complexity itself. We innately focus our awareness and attention, like a flashlight, on whatever part of a human-made or natural complex system is foremost in our mental or physical field of view. This helps us to solve problems within that field of view but comes with a disadvantage. In order to resolve either a broad issue or a specific problem that includes other parts of the complex system, we have to consider the issue/problem within that wider context. To do so, we have to move our attention around like a flashlight and then filter and combine the information we gather. We find this a difficult mental task, so we rely on simplifications and rules of thumb.

This foible is reflected in our decision making about complex systems. For example, an event in a complex system is more likely to be the result of the interactions between many variables of varying significance occurring over a range of time, space and intensity: a flexible hierarchy of causes. However, we generally assume that an effect (outcome/event) is the result of a recent, simple cause-and-effect relationship. We therefore also tend to believe that we can reliably deduce the cause(s) of an effect from a small amount of information. Our use of simplifications and rules of thumb, along with our other decision-making foibles, such as our reliance on an untrained intuition, constrains our efforts to think comprehensively and coherently about a complex system and its dynamics.

Because we under-appreciate our decision-making foibles and complexity, we tend not to compensate for either during our decision making. We tend instead to display great faith in our decisions, until we suffer the rude awakening that comes when the apparently unexpected, unanticipated or abnormal actually happens. Consider the global financial crisis of 2008. It was precipitated, in part, by the excessive risk taking of bankers and mortgage lenders and, in part, by their and the regulators' belief that there was no excessive risk taking. Their surprise when the crisis occurred was followed by a scramble to find someone to blame and efforts to retain the essence of the original flawed decisions. They set us up for another financial crisis. This completes a discussion about the foundation of our personal contribution to how we got here.

The Guide: Our Cultures

Our individual decision making does not take place in a vacuum. It occurs within the context of our group's culture, whose world view influences our preferences and decisions, and its environmental endowment. Culture's contribution to how we got here is illustrated by Western culture and its guidance to us about our relationship with our environmental endowment.

Westerners may intellectually see ourselves as participants in the dynamic, complex human-environment system, but in our day-to-day dealings with the environment, we behave as though we are remote from it. After all, our culture's world view guides us to act as environmental managers in charge of transforming the environment for our benefit rather than as participants who are trustees of its functioning for future generations. In addition, although our culture encourages us to marvel at the diversity of the earth's creatures and features, it also guides us to see and treat its structure and functioning as being, mostly, deterministic in nature, such as having a highly ordered but competitive fixed hierarchical structure. Unsurprisingly, these features are included in our view of both our own and an ideal complex culture's structure and functioning.

In keeping with that guidance, Westerners tend to treat the world as inherently describable, predictable and controllable. We imagine that we can draw reliable conclusions about it based on relatively small amounts of information. We prefer firm decisions and conclusions about our world, even though qualified judgments can more reliably represent events in complex systems.

No wonder that we tend to believe that our (intuitive or formal) predictions about the future of both human-made and natural complex systems are more valid, useful and reliable than they are. And we tend to imagine that we have a much greater ability to manage and control the functioning of the complex human-environment system than we do. In particular, Western culture guides us to prefer cures for complex problems that are simple and technological, and that satisfy our world view's emphasis on control, the notion of progress and economic growth.

These features are most clearly seen when the human-built complex systems we create, manage and participate in don't work as intended: when they fail or result in unexpected consequences and unwanted collateral effects. It is culturally acceptable, and personally convenient, for Westerners to attribute these outcomes to an uncontrollable, unforeseeable or external cause. When we do acknowledge our contribution to a failure, we prefer to treat it as an act of the incompetent. We prefer not to meaningfully acknowledge the broader, more routine aspects of our contributions to it. And we find it difficult to recognize that behind these failures lies our culturally supported, incomplete and imperfect view of our world and our relationship with it. Included in that view is our overly confident faith in our abilities. Critically, when faced with a failure, we prefer not to adjust our views and beliefs. Instead we treat these failures as prompts to improve our techniques, a position supported by our culture's notion of progress. All of these features are illustrated by our response to the flooding of New Orleans, the

Challenger's disintegration and the Deepwater Horizon oil well blowout.

These culturally primed beliefs and responses have contributed to Westerners participating in the functioning of natural complex systems as managers, not trustees. The long-term result has been the destruction and severe disruption of many ecosystems and the loss of the resources they provide to us. For example, our overfishing of the cod found on the Grand Banks off Newfoundland resulted in a crash in their population, prompting the ecosystem to switch to a low-cod quasi-stable state.[14]

In general, the guidance that the world view of a more complex culture provides to its members facilitates the provision of material resources but biases and maintains their incomplete understanding of the human-environment system. It tends to enhance the difficulty the members experience when they personally try to include features such as our foibles and complexity into their decision making. This makes it more difficult for them to decide to live and resolve issues in ways that are in keeping with the human and environmental constraints on their lives. This is culture's contribution to our current circumstance.

The Host: The Environment

The human species, like all species, is an integral part of the biosphere. We too experience its functioning and changing, rely on it to supply the resources (including services) we want and are subject to the constraints on its ability to provide them. This is a basic view of the environment's contribution to how we got here.

Details are added by considering how our relationship with our environment has changed over time. Our slow transition from simple hunter-gatherer cultures to complex industrial cultures required changes to our culture's world view. In particular, it increasingly

---

[14] Bundy, A. 2005. Structure and functioning of the eastern Scotian Shelf ecosystem before and after the collapse of groundfish stocks in the early 1990s. Canadian Journal of Fisheries and Aquatic Sciences, vol. 62, no. 7, pp. 1453–1473.

guided us to transform our environmental endowment and, in keeping with that, to see ourselves as special and largely separate from it and other species. This view of our relationship with the environment lies behind the abundance of food we produce to support an extremely large global population, our lengthened average lifespan and the availability of more material goods than our ancestors could dream of.

However, we can neither change the basic nature of how we and our environment actually function, nor remove the constraints that this places on how we can live. And although we can alter the feedbacks tying us and the environment together, we can't break them. Thus, any changes we make to our environment and how we live our lives are not costless improvements or escapes, but trade-offs. Whether we are aware of them or not, these trade-offs are also adjustments to the dynamics of the biosphere, which can change the availability of resources and how the constraints on our lives are expressed.

At some point, our efforts to transform the environment to extract resources or circumvent the human-environment system's constraints on our lives began disrupting or destroying those of its features we depend on to complete our lifeway. We were then burdened with the responsibilities of maintaining or replacing those features. In essence, as mentioned above, we have become environmental managers. We have also become responsible for ensuring that we discharge those responsibilities appropriately: we have become more intense human managers.

But, as our current circumstance indicates, we are not the most successful of environmental or human managers. We are guided by our personal world view and our culture's world view to focus more on satisfying our wants, usually using technology, than on respecting the human and environmental constraints on how we can live. We thus focus on the benefit rather than the responsibility side of the benefit-responsibility trade-off that comes with supplying and using the resources that sustain our chosen lifestyle. The environment's basic contribution to our current circumstance is the constraints that it places on our lives and the spread through the human-environment system of its response to our efforts to live as we prefer.

A Model of the Human-Environment System

The above sections discussed the contributions to our current state that originate with three complex systems: us as individuals (including our internal reality, its divergences from the external reality and our personal world view); the culture of a group of individuals (its world view with its divergences) and our environment (whose functioning supports and constrains all living things). We can more easily and fully appreciate how we arrived at our current circumstance when we treat them as the three primary factors of the complex human-environment system whose interactions form its structure and are its functioning. This section does so by integrating them into a descriptive model of the complex human-environment system and its functioning, as seen from a systems perspective (figure 2).

The relationships binding the three factors together are represented by two overlapping feedback loops: one dominantly concerned with belief (the belief loop) and the other with resources (the resource loop). The belief loop represents those belief-related relationships and interactions that bind individuals into a cultural group. An example is the member-culture feedback, in which a culture's world view guides its members' decisions, and the members' decisions change their culture's world view.

The resource loop represents those relationships and interactions that bind the individual members and their cultural group to their environmental endowment and its functioning. For example, a culture provides guidance to its members about acceptable lifestyles. To live their chosen lifestyle, the members need resources, which they secure from their environmental endowment, while following their culture's guidance about their relationship with it. The functioning of the culture's environmental endowment provides those resources and responds to their extraction, use and discarding. Its functioning can also change the endowment's state, including the availability of resources, to which the members react by following or adjusting their culture's guidance about the environment and resources. This completes the human-environment feedback.

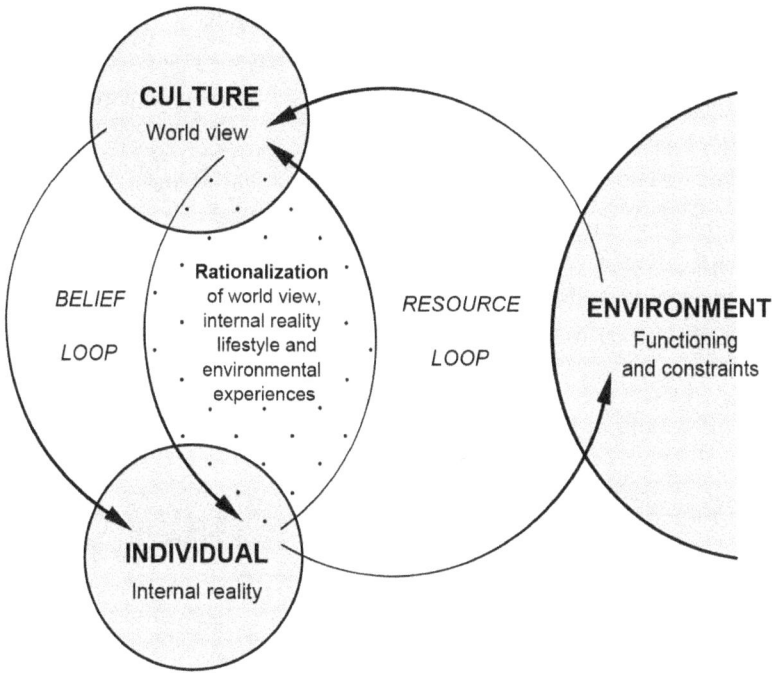

Figure 2. Our dominant influence on the biosphere's functioning transforms it into a dynamic, complex human-environment system. This is a model of it from a systems perspective. The system is formed by the interactions (including feedbacks) among three factors: our cultures (represented by their world view), individuals (represented by our internal reality) and the environment (represented by its functioning and constraints). The system's functioning is expressed by the linked belief and resource loops. In the belief loop, our culture's world view guides our decision making, while our decisions affect our culture's world view. In the resource loop, individuals, guided by their culture, use environmental resources to live their culturally influenced lifestyle and pay for their culture's reaffirmation and maintenance processes. In the process, they change the environment, which affects the supply of resources and thus their decisions. The link between the two loops is our efforts to rationalize our thinking and decisions. In essence, this means rationalizing our internal reality with our culture's world view, our experience of the world and our wants.

The belief and resource loops have features in common, such as our internal reality (including our personal world view), our culture's world view and our lifestyle. Through these common features, the loops interact with one another. The interactions are mediated by those aspects of our (culturally influenced) conscious and unconscious decision making that help us to make sense of what we know and experience (i.e., our rationalization process) (figure 2). For example, our chosen lifestyle, such as what we eat, and our explanation for it can be thought of as our way of rationalizing the (personal, cultural and environmental) influences on our lives.

Rationalization is thus a key part of the process that binds the complex workings of individuals, their culture and their environmental endowment together into a human-environment system. When seen in this context, the two interacting loops plus rationalization represent the complex human-environment system's functioning and changing: they are its dynamics and its constraints on how we can live.

Understanding How We Got Here: The Model's View

The model of the human-environment system can help us to think about our current circumstance in a more comprehensive, coherent and consistent manner, which should lead to a greater understanding of how we got here. This is illustrated by systematically using it to describe, from three perspectives, how Western culture arrived at its current circumstance.

First consider it from the perspective of the model's cultural factor. The core of Western culture's world view includes a belief in progress, economic growth, consumerism, democracy, individualism, innovation, technology and a dominion-biased relationship with the environment. Western culture thus guides us to favour these beliefs when making decisions. But we are our culture's decision makers, so our decisions can, in aggregate, change our culture's world view, completing a member-culture feedback in the belief loop (figure 2).

However, like members of all cultures, we are indoctrinated from birth to follow our culture's guidance. As a result, our day-to-day decisions tend to favour the status quo, and the core of our culture's world view displays inertia to change. In contrast, our decisions can change the periphery of our culture's world view quite quickly, as illustrated by changes in consumer trends. But the immediate influence this has on the core is limited by its inertia to change. The core of our culture almost always changes slowly, over a generation or longer. Think of changes in sexual taboos.

The belief loop overlaps and interacts with the resource loop (figure 2). In the resource loop, the beliefs favoured by Western culture's world view guide us to adopt a lifestyle that demands a significant amount of resources. Think consumerism. But the supply of the resources we want is constrained by both the effort needed to provide them (which can result in resource shortages) and by their practical finiteness (which results in resource scarcities).

Westerners are guided to preferentially resolve such resource supply issues in ways that maintain the supply rather than reduce our demand. However, the nature of resource availability means that, beyond some low level of demand, maintaining supplies requires an increasing amount of effort per unit of produced resources. We deal with this constraint by increasing our reliance on technology and applying it to more of the planet. But, in doing so, we also (directly and indirectly) increase our human and environmental management responsibilities. Collectively, the above increases favour an increase in Western culture's complexity. This is illustrated by the increase in the complexity of European culture's structure and functioning since the early 1800s that accompanied the increase in population, innovation and technology, and the associated human and environmental responsibilities. Think of the increasing need to fertilize fields using human-made nitrogen fertilizer, which began to be produced at the end of the 1800s.

All of these increases are ultimately paid for by resources extracted from the environment. By making that payment, Westerners complete the cultural-complexity–resource-demand feedback (complexity-resource feedback) in the resource loop (figure 2). The guidance from Western culture's world view ensures that this feedback will favour an ongoing increase in our demand for resources and thus an increase in our culture's complexity.

This is the case because increasing a culture's complexity involves more than just extra technology and management, and paying for the material costs. If a culture is to remain functional, then its members must change in ways that accommodate and support the increased complexity. In essence, they must (individually and collectively) make and rationalize the adjustments to their personal world view that will ensure they continue to favour their culture's now more complex structure, its increased demand for resources and the higher (material and social) costs of supporting both. And the members must believe in the cultural explanations for why they should.

In light of Western culture's features listed above, it is unsurprising that individual Westerners continue to believe that our (political and business) leaders, and our ability to innovate and apply technology, will continue to ensure that our (real and perceived) material and mental wants are satisfied, our responsibilities discharged and issues addressed. Consider that we retain our faith in the ability of technology and its material products to satisfy our psychological needs, as illustrated by our belief that increases in wealth and consumer choice lead to greater happiness. Our faith also prompts us to support the investing of time, effort and resources needed to maintain technological innovation and economic growth, and to continue, mostly willingly, to provide the labour, taxes and investment needed to pay for the increased costs of our culture's reaffirmation and authority-maintenance efforts. Ironically, these reaffirmation costs too are ultimately paid for by resources, which strengthens Westerners' complexity-resource feedback.

In summary, the guidance from Western culture's world view has ensured that, over the long term, our ancestors made, and we continue to make, the personal adjustments that collectively favour us strengthening those aspects of our culture's world view that support the increases in our culture's complexity and resource use while weakening those that don't. This is how, from the perspective of the model's cultural factor, Western culture arrived at its current circumstance.

***

By considering our current circumstance from the perspective of the model's environment factor, our appreciation for how it arose becomes more real. Western culture's world view guides us to adopt a dominion-biased relationship with our environmental endowment. Our acceptance of that guidance is reflected, in the resource loop, by how we extract and use resources, and dispose of the associated waste. As the goods surrounding us indicate, our culturally supported choices do favour us satisfying our resource wants, but they also (knowingly and unknowingly) increasingly disrupt the environment's functioning and thus its ability to provide them. Westerners therefore inevitably face environmental issues and acquire environmental responsibilities. We are guided by our dominion-biased relationship with the environment to discharge them by becoming environmental managers, which reinforces our dominion-biased relationship with our environment. This completes a feedback in the resource loop between the environment's state and us: a manager-environment feedback.

Our dominion view of our relationship with our environment and our faith in the notion of progress both reinforce our culturally guided preference to resolve environmental issues by using technology. In doing so, we invariably circumvent the environmental constraints on how we can live, rather than change our goals and behaviours to live within them. But circumventing these constraints doesn't neutralize or remove them; rather, it transforms or heightens their expression. To overcome these additional or heightened constraints, we apply technology, thus completing an environment-technology feedback. It also reinforces our dominion-biased relationship with the environment and favours an increase in our environmental management responsibilities.

There is a population dimension to this discussion. In the past, Western culture's world view favoured large families. Today it favours a family unit of no more than four members (two adults, two children). However, it still has little to say about the ideal absolute size of its population. If anything, Western culture favours an increasing population because, so economists tell us, it helps to maintain economic growth. These views join with the other features of Western culture's guidance, such as its dominion relationship with the environment, to ensure that our decision making largely ignores the other medium- to longer-term results of population growth, such as increasing resource usage, and the limits to population growth, such as social issues and the constraints on the supply of resources.

The overall result of these features and feedbacks is an ongoing absolute increase in our use of resources, our increasing disruption of the environment and thus a corresponding increase in our environmental responsibilities. But, as the record shows, we are poor environmental managers. This is how, from the perspective of the model's environmental factor, Western culture arrived at its current circumstance.

***

Next, consider the perspective of the individual factor. The resources Westerners secure are the material means by which we maintain our culture's complex structure, its dynamic functioning and its world view. In contrast, our decisions are the mental means by which we maintain or adjust our culture's world view and our resource wants and how they are satisfied.

At all levels, elite and public, of our culture's power hierarchy (its structure), individual Westerners are routinely making these day-to-day and executive decisions (institutions can't really make decisions). They are the product of our decision-making foibles and the influences on them. Those influences include our internal reality (with its divergences); other members; our culture's world view (with its divergences); and our experiences of our environment. All aspects of our decision making are linked together and interact with one another (including feedback) to form a process in the human-environment system represented by rationalization (figure 2). With this in mind, a discussion about Westerners' choice of lifestyle illustrates how the individual factor contributes to our current circumstance.

In the belief loop, Western culture's guidance, such as favouring the notion of progress, a focus on economic growth and material wealth, prompts us to choose a resource-intensive lifestyle. Think of the stuff you own and desire. In the resource loop, this choice contributes significantly to the type and amount of resources our culture uses (figure 2). It also raises questions, such as: How should we acquire these resources? How should we dispose of, into the environment, the waste from acquiring and using them? How should we deal with the associated disruption of the environment? And when does consumption become over-consumption? As expected, how we answer these questions is influenced by our culture's world view, such as its dominion-biased view of our relationship with our environmental endowment and, of course, its focus on economic growth.

Our answers to these questions, and our choice of a lifestyle, are also influenced by more personal, less material concerns. Consider that we prefer choices and answers that provide us with the feelings (sense) of stability, order, belonging and meaning that we innately desire. These various kinds of influences are not isolated from one another: they are interconnected and interact. An example is our tendency to (usually unconsciously) equate an increase in our sense of social-psychological well-being, such as happiness, with an increase in our material wealth.

Collectively, these influences favour us aligning our choice of a lifestyle and answers to the above questions with the guidance from our culture's world view. In doing so, our culture's world view becomes more real for us, which favours us following its guidance, thus completing a feedback. The overall result is our tendency to make decisions that favour us reaching for current cultural goals and maintaining current social and environmental trends (business as usual).

There is more. The presence of divergences in both our internal reality and our culture's world view combine with our decision-making foibles and the world's complexity (think feedbacks) to allow us to make decisions that we feel are socially and environmentally appropriate but are, in the longer term, actually at odds with the external reality. Our decisions, such as our choice of lifestyle and answers to the above questions, can thus take us and our culture's world view outside of the human and environmental constraints within which we must live, if we wish ourselves and our culture to remain functional over the long term (i.e., become viable). Our decisions can also hold us there. We find it difficult to recognize when we have made such decisions and what lies behind them. We struggle to decide on and implement the (mental and material) changes needed to avoid repeating that type of decision and undoing the outcome.

Consider what happens when we, individually, become personally aware that some aspect of our personal lifestyle or our culture's average lifestyle is negatively affecting either us, our community or our supporting environment. Think of our cars' exhaust warming the planet, our plastic waste polluting the oceans, the environmental consequences of a diet rich in red meat and the health consequences of a diet rich in highly processed foods. At this time, we may experience the shock of cognitive dissonance or a reality paradox.

If we do choose to pay attention to these impacts and the issues behind them, then we preferentially use a social or cultural context and reference to judge them. And we tend to (consciously or unconsciously) look for explanations and solutions within our culture's resource loop (figure 2). In particular, we look for technical, resource and managerial solutions that satisfy our culture's notion of progress, economic growth and a dominion-biased view of the environment. Think of efficiency increases, laws, innovation, environmental transformation (e.g., dams) and material-dominated (e.g., technology-based) environmental restoration. This focus makes sense because these types of solutions are easier for us to rationalize and implement. After all, they are consistent with Western culture's guidance about the preferred ways of resolving issues, and they maintain business as usual. At a personal level, they are emotionally benign and maintain our current lifestyle. Think of resolving climate warming by focusing on carbon capture and green energy sources. We can also believe, or be easily persuaded, that any "unexpected" environmental and human impacts resulting from a particular (usually technological) solution can be reasonably addressed by applying another such solution.

If, instead, we use a personal context and reference to pay attention to these impacts and the issues behind them, then we prompt ourselves to look at our personal contribution. However, the majority of us, whether in our role as scientist, capitalist, politician, recreationist, environmentalist, farmer, priest or parent, or just a member of the public, prefer not to look too deeply into the mirror. Instead, we (consciously or unconsciously) seek out masking and self-satisfying excuses, or alternate explanations, that let us personally off the hook, such as pointing the finger at others or discounting our own contribution. We can also displace our attention away from ourselves by redirecting our efforts to a search for solutions within the resource loop, as discussed above.

In both contexts, we prefer to avoid any prompts to pay close attention to the belief loop. This tactic makes sense if we consider our desire to avoid the discomfort that can arise when looking in the mirror or going against the guidance from our culture's world view. After all, searching for a belief loop solution would require us to scrutinize our personal and cultural beliefs, preferences and values, our divergences, our decision-making foibles, and our limited degree of awareness or attention to our surroundings and ourselves. And any such solution would likely impinge upon our sense of self and the essence of our culture's world view.

Our avoidance of the belief loop is a significant oversight because our personal and cultural beliefs and values are an important source of the guidance we rely on during decision making, including those decisions that affect ourselves and our environment. For example, by considering belief loop solutions, we are more likely to recognize when our personal lifestyle is out of step with environmental reality. We are also more likely to accept that behind that critical divergence lies our decision-making foibles and the influences on them. Think too, in practical terms, of advertising and social media. We are then in a position to make and implement decisions that would reduce those divergences or, at least, their influence.

But we usually choose not to. Thus, helped by our cultural indoctrination, perceptual limitations and decision-making foibles, our divergences persist, and we continue to make decisions, such as lifestyle choices, that discount the human and environmental constraints on how we can live. This completes the discussion of how, from the perspective of the model's individual factor, Western culture arrived at its current circumstance.

Collectively, these three perspectives describe how Western culture came to be in its current situation.

***

In summary, chapter 3 described a model of the complex human-environment system that is based on its three primary factors: individuals, their culture and the environment. Using it helps us to appreciate how Western culture, an example of a complex industrial culture, arrived at its current circumstance.

Most of the world's population are members of its more complex industrial and industrializing cultures; and all cultures inescapably participate in the human-environment system. Therefore, the model and the essence of the explanation given for Western culture's circumstance can provide us with a broad overview of how we arrived at our global circumstance.

A large, growing global population of humans, their increasing resource wants (including services) and the environmental consequences of securing, using and disposing of them are our material contribution to our current circumstance. However, behind that material contribution lies our mental contribution: our personal, culturally influenced beliefs, preferences and decisions.

When making our conscious decisions we, personally, experience difficulty taking our decision-making foibles, and the personal and cultural influences on them, into account. We also experience difficulty appreciating and accommodating complexity. We therefore tend not to recognize that our decisions have resulted in us becoming environmental and human managers. Or that we, as members of a more complex culture, are not particularly good at either. In particular, we tend not to fully consider the human and environmental constraints on how we can live.

The overall outcome from the combination of our decision-making foibles, the guidance from our culture's world view and the environment's response to our decisions is our current global circumstance.

Yes, humans are smart and powerful. Our current global circumstance did not arise because we are evil or stupid, but because we are well and truly out of our intellectual and emotional depth. Fortunately, just as the model can help us appreciate how we arrived here, it can help us make sense of the diversity of opinions about what the future may hold for us, what we may want to do about it and what we may actually do.

Chapter 4

WHAT THE FUTURE HOLDS

We make predictions about the future for a number of reasons: to reassure ourselves, to accept it or even to manage it. This chapter discusses predictions about our and our environment's possible futures.

The Biosphere's Likely Future

We are both part of and supported by the earth's dynamic, complex biosphere. Currently, our influence on its functioning is so pervasive that, at a human-relevant scale, it is now a human-environment system. The biosphere's future will therefore be significantly determined by how our current and future culturally influenced decisions will affect its functioning: that is, the degree to which the cumulative effect of our growing population's demand for resources, disposal of waste and environmental manipulation will disrupt the system physically, chemically, biologically and ecologically. The current trend in those impacts could continue for a while because we are preconditioned to exert our considerable influence on the environment until we can't, rather than proactively refrain from doing so.

If we do maintain this trend, then the earth's atmosphere is expected to continue its rapid warming and will likely exceed our <2.0°C warming target. That exceedance will contribute significantly to further environmental changes, including sea level rise, ocean acidification, species extinctions and biome transformations. However, even if warming stays below this target, the cumulative impacts of our ongoing activities, such as land clearing and the distribution of exotic species and genes, will still result in significant environmental changes. We can expect that most larger, specialist and less adaptable species will become extinct, the populations of most other species will continue to rapidly decrease and novel ecosystems dominated by weedier species will appear.

Consider too that even if we were to stop disrupting the biosphere today, its inertia insures that the current transformations will continue well into at least the medium term. One way or another, it seems that we, but especially the younger generations, will likely become earnest, noisy bystanders, wringing our hands as we experience the human-disrupted biosphere slowly being transformed into an increasingly hot, stormy and ecologically simplified state.

If we wish the biosphere's future to be different, then we face some tough decisions. However, before making them, we should remember that there are uncertainties in the above predictions. Consider that we favour predictions in which the future is a continuation of the current inertia-influenced trends we prefer. But, the presence of chaos, randomness and our decision-making foibles ensures that deviations from those trends are inevitable, and they are more likely to occur the further into the future we predict. Examples of factors that could contribute to an alternative environmental future are the following. There appears to be insufficient fossil fuels to support at least the most dire predictions about our climate future. However, we may have already initiated an abrupt (self-sustaining) warming event. And the possibility of significant trend-disrupting social unrest is real.

In overview, it seems that up until at least the medium-term future, we are unlikely to have voluntarily changed our world view, lifestyle or population in a sufficiently appropriate and timely manner to deviate from current trends. Thus, at least in the medium term, it is unlikely that we will be living within the environment's constraints. The current level to which the biosphere's functioning has been disrupted suggests that, in the longer-term future, any deviations from the environment's current trend will favour greater environmental transformation, not less. Other possible futures will be discussed below.

Satisfying Our Future Resource Wants

Although our future wants for each resource (including services) will be different, the nature of their future supplies will still display some common features. Critically, if the growth rate in our population and our demand for water, materials, food and fuel continue on their current multi-year trends in the 1% to 3% per year range, then they would experience a doubling within 70 to 24 years, respectively. But, in a finite world where resource renewal rates are limited, these growth rates cannot continue for long before resource shortages, then scarcities, arise.

Consider too that our resource future will arrive as part of the dynamic, complex human-environment system, so we should contemplate it in that context. Unfortunately, our personal efforts to do so are hindered by the constraints and biases affecting our thinking about resources. For example, we tend to preferentially focus on those features of a resource that we know something about and consider to be of immediate relevance to our personal day-to-day lives. If we do recognize the connections among resources and our demand for them, then our broader contemplation of a resource's future is constrained by the difficulty we experience comprehensively taking the system's complexity into account. For example, we still tend to focus on the simpler relationships and the short-term, more obvious changes. We tend to miss the more basic, often subtle feedbacks and their longer-term influences on our resource future. And we tend not to appreciate our day-to-day contribution to that future.

This is illustrated by our contemplation of our food future. We tend to focus on the amount of food we might want in the short term and our ability to produce it. In this context, it is easier to pay attention to, for example, the present state of the soil in our fields, particularly its fertility, rather than its slowly developing longer-term state, such as the buildup of excess nitrogen and the loss of soil carbon. We also pay limited attention to how farming indirectly affects the production of food. For example, fertilizing fields with human-made fertilizer impacts the production of food from the oceans, while farming's emissions of greenhouse gases contributes to the change in climate, which affects agriculture. Consider too the feedback between what we eat and what we grow. For example, a diet rich in red meat contributes to climate warming, which affects the growing of the feed used to raise the animals that provide the meat.

A broader view of our food future would require us to coherently imagine how enough food will be produced to feed a growing global population during a time when climate warming is affecting the practice of agriculture on many fronts, from changing freshwater availability to rising sea levels inundating the land. Then add in the effect that changes in the supplies of oil (think fuel) and natural gas (think fertilizer) will have on industrial agriculture and the importing/exporting of food. Finally, consider that a change in energy availability also affects the ability of an industrial culture to maintain its world view, and thus its guidance about resources, including the food we should eat and how to secure it. Treating this as a coherent whole is difficult for us.

In overview, we tend not to realistically consider the constraints on the supply and production of resources that arise from the complexity of the human-environment system. Thus our ability to alter the earth in order to secure the resources we want today is far ahead of our ability to assess the effect that our earth-altering actions will have on the supply of those resources tomorrow. Instead, we tend to grasp, and hold onto, simplistic predictions about the future that match both the guidance from our culture's world view about the future and our innate, short-term, usually optimistic outlook. Westerners, for example, prefer to focus on the next technological breakthrough rather than a change in beliefs or values.

Some of these shortcomings in our thinking can be addressed by comprehensive global resource models that take into account the connections among resources and represent our possible choices about how to live as scenarios. For example, their projection of the business-as-usual scenario suggests that, in the future, our increasing demand for resources will continue to be met, but with increasing difficulty and an increasing impact on our ability to supply resources in the future (overshoot). Eventually our ability to provide them will be exceeded and the supply will not just decline but collapse: overshoot and collapse. This will be followed by a rapid decline in our population. However, there are a number of difficulties with the conclusions from these global models. They lump all cultures and resources together when, in reality, the earth's resources are unevenly distributed and its cultures diverse. And their use of scenarios treats our choices as fixed, meaning that the model applies that choice throughout the time period of its projected future. In reality, we respond to events dynamically: we can change our choices over time.

Fortunately, by combining the models' broad conclusions with both knowledge of our (personal and cultural) foibles and studies of specific resources, we can draw a more realistic conclusion about our resource future. Both the ongoing trends in our use of resources and a retrospective review of one model's projections indicates that we are indeed following a business-as-usual path into the future.[15] However, the future along this path is unlikely to arrive in the simple way the models suggest. As the future arrives, each culture will face a different aspect of a particular resource shortage, to which its members will be individually guided to respond by their personal wants and their culture's world view. Their basic choices include paying more to secure the resource, changing their preferences (lifestyle) and migration. The aggregate of the members' responses becomes their culture's response to the shortage/scarcity: a mix of voluntarily changing its resource preferences, increasing its efforts to maintain the supply and applying technical innovation. These personal and cultural responses can change over time.

Rather than the single event suggested by the models, our global resource future will likely arrive as a patchwork of different resource shortages, scarcities and cultural reactions to them that will be spread over time and space. But, it is also true that today's cultures are no

---

[15] Turner, G. 2008. A comparison of the limits to growth with thirty years of reality. Commonwealth Scientific and Industrial Research Organization (Australia). Socio-economics and the environment in discussion, CSIRO working paper series 2008–09, pp. 1–49.

longer largely stand-alone, self-sufficient entities. Today, economic globalization and climate warming, as two examples, have tied us all tightly together. It is also reasonable to expect that the members (public and elite) of the more complex cultures are likely to continue exerting their considerable influence on the environment in an effort to satisfy their current large resource wants (maintain business as usual) until they can't, rather than proactively (voluntarily) deciding not to. Think overshoot. Thus, it is indeed possible that some combination of events could affect so many countries that it results in a global-scale resource supply shortage, scarcity and then a collapse. Think of the global financial crisis of 2008 and the related food issues, or the early 1970s and late 2000s oil crises.

For a particular country, the final outcome of its reaction to a shortage/scarcity of a resource will depend on a mix of its human-environment system's features. Consider that the influence of a culture's world view on its members' decisions, such as its preferred method of resolving issues, will be joined by other factors: the members' expectations; the level of inequality in the distribution of wealth and access to resources; and the state of its environmental endowment. For example, the more unequal the wealth distribution, the greater the population density and the higher the people's unfulfilled expectations, then the more likely it is that resource shortages will result in social instability. Demographics also matter. For example, if the largest proportion of a country's dissatisfied population is under 30 rather than over 50, then it is more likely to face social and political unrest before it experiences a primarily resource-induced cultural "collapse."

However, some argue that these predictions deal with the future, and thus can't be proved by measurements or observations today. Therefore, they are mere speculations, and any action based on them is a waste of time. And anyway, the argument continues, we have muddled through before, so we will muddle through again, especially with our superior technology. This perspective is appealing because it fits with our innate optimism and the positive illusions we tend to hold about the future. It is particularly appealing for Westerners: it matches our culture's world view.

For example, Westerners honestly believe that the issues of climate warming, peak conventional oil and gas, and long-term food security can all be simultaneously resolved for the long term by developing non-fossil-fuel energy sources. We also tend to believe that these goals can be reached without significantly changing our lifestyle, our culture's world view and our environment's functioning, even as the global population grows. Certainly, holding an optimistic outlook provides us with the positive boost that we need if we are to deal with the vagaries of life and muddle through to the future. But what if holding onto those hopes takes us to a place we never intended or wanted to go?

Fortunately, history provides us with an opportunity to verify these arguments and counter-arguments about the nature of our future and how to address it. It also provides us with other views of the future and possible solutions.

Lessons from the Past

History provides us with a longer-term overview of the (internal and external) influences on a culture and how the members responded to them. It highlights the slower changes while providing the framework or context for the faster changes. As previously discussed, the historical record reveals that, over time, some simpler and less complex cultures experienced an increase in their level of structural complexity. It also shows that some complex historical cultures experienced a decrease in their complexity (a resimplification), which is the topic of this section.

Just as your perception of an aspect of a complex system is influenced by the perspective you take, especially the scale (of time, space and intensity) you focus on, so is your perception of a particular historical resimplification. Your perception is also biased by your personal and your culture's world view. For example, Westerners refer to the resimplification of what we consider to be history's iconic complex cultures, such as the ancient Maya and ancient Romans, as a cultural collapse. This expresses our tendency to treat resimplification as both a fast event and a warning to ourselves, a failure for us to fear. To fully appreciate a historical resimplification, you need to take these influences into account. The same is true of your ponderings about the possibility of a present culture, including your own, undergoing a future resimplification.

With this in mind, each historical resimplification was unique in detail, but all arose from a mix of the same broad features, such as the culture's level of complexity, the state of its environmental endowment, its members' culturally influenced response to events and its wider social/political context. For example, we tend to think that the Greenland Norse disappeared because of a cooling climate, but there were other contributions. Their environmental endowment (e.g., thin soil, few trees and a short growing season) was marginally suited to their type of mixed farming that used European livestock. But their culture's world view, such as its strong sense of community, allowed them to adapt to these conditions. For example, they hunted seals to feed themselves during the spring hard times. To secure resources that were unavailable to them in Greenland they traded prestige items, such as narwhal tusks, with the mainland Norse. However, as the climate in Greenland became cooler and more unsettled at a time when they were operating their farms at maximum productivity and their trade items had fallen out of favour, they found themselves living outside of their culture's functional boundary. Their culture's world view now limited their ability to adapt further.[16] The resimplification of the Mayan, Roman and Easter Island cultures similarly involved a number of interacting aspects of their respective human-environment system.

Resimplification is thus not, as we tend to assume, the result of one or two factors, such as a resource scarcity, climate change or social disorder. Nor is the process of resimplification as smooth as the

---

[16] Dugmore, A. J., McGovern, T. H., Vésteinsson, O., Arneborg, J., Streetera, R. and Keller, C. 2012. Cultural adaptation, compounding vulnerabilities and conjunctures in Norse Greenland. Proceedings of the National Academy of Sciences, vol. 109, no. 10, pp. 3658–3663.

output from the comprehensive global mathematical models indicates. Indeed, resimplification is the outcome of a messy, complex mix of many factors that interact and change at different rates and intensities over an extended period. It is the outcome of a process that can be described as a flexible hierarchy of weak to strong, local to regional/global, and short- to longer-term interactions that combine into trends and changes that favour (preconditioned) resimplification.

The record also shows that some historical cultures could recover, temporarily, from a resimplification. They could even repeatedly oscillate between a complexity increase and resimplification. Babylonia's long history is an example. Thus, some cultures existed well beyond their theoretical "collapse" date. However, eventually, all of the more complex historical cultures (larger states and empires) experienced resimplification to such a degree that they either disappeared or became disconnected from the complexity of their prior selves. Thus, beyond some level of complexity, becoming more complex is counterproductive: it favours future cultural resimplification.

Our appreciation for a historical culture's process of resimplification increases when we apply the descriptive model of the human-environment system. We can then conclude, for example, that during a historical culture's resimplification, the features the members considered, their interpretation of those features and their chosen response were likely all influenced by the same basic factors mentioned above for us today. They include the guidance from their personal and their culture's world view, their choice of perspective from which to look at their circumstance, and their decision-making foibles. In particular, the members likely focused on those events and trends that were stronger and of immediate, direct human interest, such as a battle. They were likely unaware of most, if not all, of the weaker or slower trends, regardless of their potential significance. Think salination for the Babylonians. Their response to the trends and events they did see would likely have reflected what their culture thought was appropriate, regardless of its relevance or effectiveness. To make this clear, consider which of today's events those Christians who believe in the apocalypse and rapture, and those capitalist/ neoliberal economists with their strong belief in wealth and growth, tend to focus on and how they interpret those events. What do Westerners in general focus on?

When the model includes the historical data, it helps us to appreciate how past and present cultures may, within limits, delay a resimplification, partly recover from one and even experience more than one. Simplistically, it depends on how a culture's world view guides its members to respond to an adverse event, in conjunction with the ability of the culture's supporting environment to recover from its disruption.

The combination of historical data and the model also helps us to make sense of what in our short-term, close-up view of today's complex cultures appears to be a chaotic jumble of events and influences on their future (figure 2). This helps us to determine whether and how today's more complex cultures are likely to resimplify.

Consider that their world views all tend to guide their members to prefer solutions to day-to-day issues that maintain or increase their culture's resource use, complexity and responsibilities. At the same time, these world views tend to limit, not enhance, their members' awareness of both the constraints on their culture being functional over the long term and the indications that they are failing to respect those constraints. The members (elite and public) thus tend to have a biased and incomplete view of their current circumstance. In particular, they fail to fully acknowledge their social and environmental responsibilities. When they do become aware of those responsibilities, they tend to exhibit a culturally supported inertia to appropriately addressing them. After all, that would require the members to each adjust their lifestyles and personal world views. They are thus likely to either fail to discharge their responsibilities, or fail to do so on the external reality's terms. Instead, their decisions tend to favour the essence of the status quo. Collectively, these types of features help to ensure that today's more complex cultures are, like those in the past, inherently unstable and liable to resimplify. And their culture's guidance ensures that the members will face challenges when making and implementing the decisions needed to avoid it.

If today's more complex cultures do continue on their current path, then many will likely go through a resimplification process. During it, they will experience the features common to all resimplified cultures, but they will each express those experiences in their own unique way. Those complex cultures that include a complexity-resource feedback or equivalent (and thus a greater inertia to reversing direction) are the most likely to resimplify. But this does not mean that the other complex cultures are immune. After all, today's cultures are linked together through globalization and a common, changing environment. It is thus possible for some cultural resimplifications to occur as part of a multicultural event.

History and the model also allow us to outline the experience of resimplification. As a historical complex culture neared the limit to its functional level of complexity, its members would experience an increasing number of trends and events that hinted at resimplification, such as diminishing resource returns for their efforts, resource supply limits, social dysfunction and difficulties appropriately fulfilling their responsibilities. The members would be increasingly forced to face resource, environmental, individual and cultural constraints and issues that they couldn't circumvent: they could no longer maintain their culture's level of complexity. The culture had to either stabilize or resimplify. But, the presence of inertia favours a culture, especially a more complex one, striving to stay on its preferred path (business as usual) for as long as possible: that is, until it is forced to resimplify.

Resimplification occurs when the members can no longer keep most of their resource, social, environmental, individual and cultural responsibility balls in the air. We, however, are strongly inclined to blame resimplification on the dropping of a particular ball, which we label as the trigger event or cause. But it is more realistic to treat a particularly notable event as just a marker on the culture's longer, more complex journey to resimplification, a process which started well before any balls were dropped. After all, the nature and speed of a culture's resimplification depends more on the state and dynamics of the human-environment system and the members' culturally influenced decisions over time than on any one event. Similarly, the fate of a culture after resimplification depends on a number of factors: its remaining environmental endowment, cultural resources, how the awareness of the members (public and elite) has changed, and the changes they are willing to make to their personal and their culture's world view.

Possible Futures

We make predictions about the future to help us prepare for it. This section divides common Western predictions about the future of our complex industrial culture into a few generic types, provides thumbnail descriptions of each and evaluates them. Because the members of industrial and industrializing cultures make up the vast majority of the world's population, this discussion reflects predictions about our possible average global future (i.e., when disregarding regional-scale cultural and environmental differences.)

There is an optimistic business-as-usual prediction that Westerners will muddle through to a satisfactory, even somewhat utopian, future. In this prediction, when social and environmental issues arrive, we will apply innovation and technology that will successfully maintain the status quo. This is what Westerners usually mean by *adapt* and *mitigate*. On the surface, and based on the last 150 years, this prediction sounds realistic. It is certainly the most likely to be acted upon, because it requires us to simply continue having faith in our culture's world view (e.g., the notion of progress, dominion over the environment, efficiency, economic growth, technology, ingenuity, etc.).

But, to favour this "muddle through" prediction implicitly requires us to accept an illusory view of ourselves and our world view. We must discount or reject our knowledge about the biases in our decision making and the presence of divergences from the external reality in both our internal reality and our culture's world view. We would also have to reject the lessons of history and our ongoing record as poor environmental managers. To act upon this prediction, we would have to feel sufficiently confident that the future won't label us "those idiots of the past."

A version of the optimistic business-as-usual prediction suggests that "we can go it alone." This prediction relies heavily on the arguments that we have the ability to look after ourselves and the environment doesn't care, in an emotional sense, about its future. Our impacts on it are just another aspect of its ongoing change. Thus, we are free to focus intently on applying innovation and technology to maintain business as usual, regardless of our impacts on our environment. Technology will solve our problems.

To favour this prediction requires that we feel comfortable abandoning any obligations to present generations of species who may fall by the wayside. We must also not only accept the increase in our disruption and destruction of the environmental functions that provide the resources (including services) that we and our culture depend on, but embrace our resulting greater responsibility for maintaining or replacing those functions. And we would have to believe that we can appropriately discharge those responsibilities. Adopting this prediction would thus require us to deny the strong evidence to the contrary: we have had limited success at recognizing, accepting and appropriately fulfilling our current responsibilities; and we do indeed rely heavily on our environment to self-manage its functioning.

There is also a pessimistic/fatalistic business-as-usual prediction. It recognizes that our human characteristics and the inherent instability of the more complex cultures favour resimplification. In this prediction, resimplification is inevitable because, when faced with critical decisions, we will either invariably make choices that drive the culture beyond some critical limit or fail to react in a timely and appropriate manner to signs of destabilization, or both. A common explanation for this point of view is that our dominant biological and mental characteristics are those of our distant hunter-gatherer ancestors, which are ill-matched to run a complex culture, especially an industrial one. Another explanation is that the elite/ leaders are invariably corrupt/incompetent. The conclusion from this argument is that all we can do is feel comfortable just carrying on, business as usual, while expecting an undesirable outcome.

But there are significant practical and emotional downsides to favouring this prediction. To hold it, we have to severely discount or even reject any claims to advanced knowledge, a powerful gift of consciousness, modernity and technology. An individual has to personally accept a severe blow to their sense of self-worth and pride in belonging to a complex culture. We also have to give up hope.

In contrast, there is an optimistic alternative future prediction in which we successfully change our current business-as-usual trends. It recognizes that resimplification looms but believes that we have the capacity to avoid it. It imagines us becoming actively engaged in implementing a mix of personal, cultural and technological changes that will allow us to quickly adjust ourselves and our cultures' world views in ways that will also maintain both the environment's functioning and our current living standard. Typically, attention is focused on a particular current social or environmental issue.

To favour this more optimistic prediction means accepting an often idealistic/simplistic view of technology and how we (as individuals) and our cultures function and change. The suggested solutions and the predicted outcomes from implementing them tend to be unrealistically straightforward or treated in isolation. An example is the belief that our energy and related environmental issues (think climate warming) can be quickly resolved by us proactively promoting the generation of energy from clean energy sources (e.g., alternative and fusion energy) and an increase in efficiency. As the belief goes, if we use this clean/green energy to power our new energy efficient lifestyle, then our lives will be improved and the environment rescued: resimplification will be avoided. This optimistic prediction, like the other optimistic predictions, has overtones of a utopian future.

When contemplating the future, Westerners also pay attention to global-scale mathematical models of our complex human-environment system that explore its possible futures and possible solutions. The conclusions from these models are quite general and mostly focused on material and population changes. Think of overshoot and collapse. In particular, they largely ignore or simplify our decision-making foibles and the nuances of personal and cultural change. Acting on solutions that are based heavily on these models will likely change the current business-as-usual trends, but they are unlikely to result in us avoiding resimplification, unless both the models', and our, limitations are taken into account and the relevant details added in.

There are variations to the above predictions, such as a more or less religious or scientific focus, but they don't change the essence of the following conclusion. All of these predictions provide us with some insight into the possible future of not just Western culture but, at least, the other industrial as well as the industrializing complex cultures. However, as pointed out, these predictions all suffer from limitations and include features that are at odds with the external reality. They can all be described as at least incomplete. For example, the further into the future they look, the greater should be the uncertainty they attached to their predictions. This is often not the case.

Is there a more comprehensive outlook for the future that includes a more realistic view of how we, especially the members of industrial and industrializing cultures, might react to its arrival? The descriptive model of the human-environment system and history help us to provide one. It is unrealistic to expect us to make reliable, detailed predictions about the future of Western industrial culture, or any other industrial or industrializing culture for that matter. However, when the current, culturally supported global trends in resource usage and environmental impacts are used in the descriptive model, it indicates that, for at least the medium-term future, our cultures are likely to face significant environmental and associated resource and social issues. Our preferred efforts to deal with these issues will also prompt an environmental and social response. That will, in turn, affect our way of life, to which we will react. This completes a feedback loop. In it, our preferential response to issues will, as discussed, tend to maintain the current or equivalent resimplification-favouring trends, even as the details of our culture's world view change. If maintained over the long term, these trends will result in the resimplification of at least the inherently unstable more complex cultures.

The scale and speed of our extraction and use of resources, disposal of waste and impact on the environment indicated by these trends joins with the biosphere's inertia to ensure that there is limited time for us to react, if we wish our culture to avoid resimplification. To take advantage of this window, we would have to immediately expend the effort needed to change ourselves and our decisions in ways that better recognize the human and environmental constraints on how we can live. But, considering the characteristics of personal and cultural change, such as inertia, preconditioning, and our preference for the status quo, the amount of time available to do so is likely too short. It is more reasonable to predict that, over the longer term, at least Western industrial culture will experience a degree of resimplification.

Even so, the members of industrial and industrializing cultures can still choose the type of resimplification we will experience. We could decide to continue following current trends that indicate a forced resimplification. Or we could slow or reverse the trends in a manner and to a degree that will at least reduce the worst effects of forced resimplification: we could aim for at least a voluntary resimplification. That opens the door for a realistic alternative future to business as usual.

***

In summary, chapter 4 suggested that the track the human-environment system is following will likely result in many of its human cultures experiencing a forced resimplification. Human and environmental inertia to change ensures that it is unlikely that these resimplifications can be totally avoided. However, it is possible for us to reduce the worst effects of a forced resimplification by deciding to aim for a voluntary resimplification.

Chapter 5

AN ALTERNATIVE FUTURE

By choosing the goal of voluntary resimplification, we give ourselves the option of working toward an alternative future that would both reduce the impacts of a forced resimplification and increase the likelihood that we and our culture will become and remain functional over the long term (viable). Indeed, reaching for voluntary resimplification is more likely to lead to viability than other efforts to avoid forced resimplification. Think of the focus and motivation associated with trying to win a war rather than trying to avoid losing it. This chapter discusses viability as a realistic alternative future for all cultures to strive for. It also discusses, from four perspectives, the challenges we face striving for it, and thus for the goal of voluntary resimplification.

The Option of Viability

Humans live in groups, each of which creates a culture. The group we belong to and its supporting environment form a human-environment system. If it is to become functional and remain so over the long term (i.e., viable), then our personal and our culture's world view must guide us to live within viability's human and environmental constraints. The feedbacks in the human-environment system (e.g., the member-culture feedback) ensure that these constraints interact with one another to collectively form a flexible viability boundary around the system (figure 3).

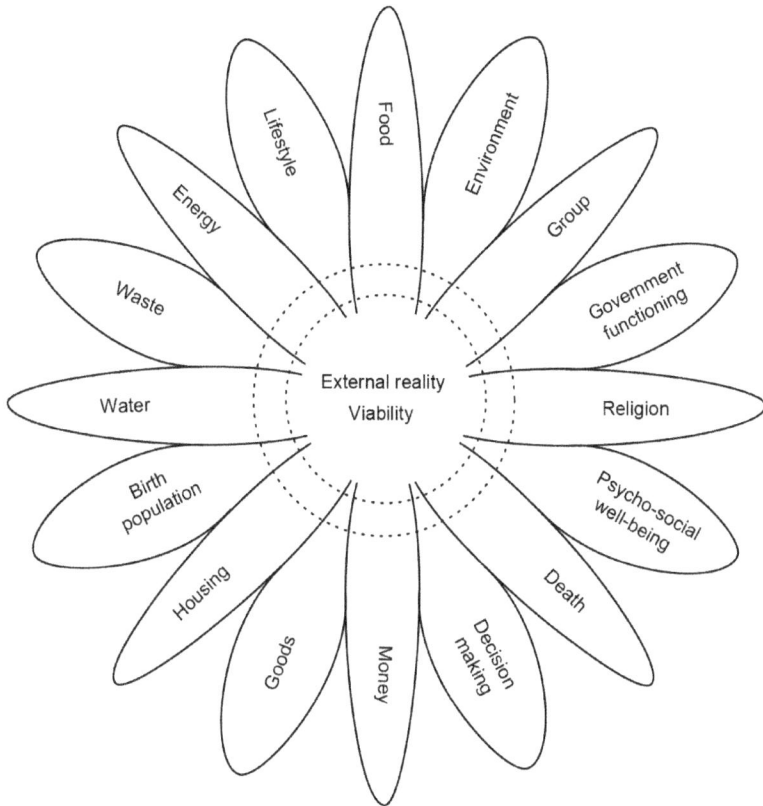

Figure 3. The viability flower: a human-centred perspective of the human-environment system. Each petal of the flower represents our personal or our culture's view of an aspect of the human-environment system that is important to us as a social group and to each of us as an individual, a member of a culture and as a member of an ecosystem. Collectively, the flower's petals (i.e., the flower) represent our personal or our culture's world view. The centre area of the flower, inside the dashed ring, represents the external-reality features of each aspect/petal. The external-reality view and our internal-reality view thus overlap in the centre area. The further outside the ring our view of an aspect falls, the greater the divergence between it and the external reality, and the harder it is for our personal, or our culture's, world view to be coherent and consistent. The boundary between the external reality and our internal reality is marked as a ring rather than a line because the boundary is gradational and somewhat flexible (fuzzy), not sharp and fixed. The merging of the petals within the centre also represents the dynamic interconnections, like rationalization, and feedbacks that link these aspects together into the single human-environment system, as presented in figure 2. The centre and its boundary thus also represent viability and its boundary, which is a complex mix of viability's constraints. Sustainability/sustainable development could be represented in the figure by a petal.

Here are somewhat idealized examples of viability's human and environmental constraints relevant to each of the human-environment system's three primary factors (individual, culture and environment). An individual should be raised in a social and physical environment in which they experience few adverse childhood experiences, eat a healthy diet and engage in sufficient exercise.[17] An individual's living conditions should also provide them with feelings of belonging, being respected and being treated fairly: they should feel safe. The way these interconnected conditions are satisfied during their upbringing should also guide individuals to develop a personal world view and a sense of purpose that favour them satisfying viability's conditions.

At the same time, their culture's world view should guide its members to favour them being trustworthy, fair and co-operative as their default behaviour. It should also provide guidance about when (and when not) to engage in self-interest, competition and anger, and how to do so in ways that favour viability. Their culture's world view should inform the members of both their responsibilities, such as satisfying viability's conditions, and their rights and freedoms. In complex cultures, it should guide the members to restrain the influence of the elite, especially those with wealth or in positions of power.

An individual's upbringing and their culture's world view should, together, guide the members to respect and protect the functioning of their culture's environmental endowment. After all, it provides the members with the resources (including services), such as fresh water, air, food and energy, that support their lifestyle, their personal world view and their culture's world view. To achieve this, the members are obliged to adopt a relationship with their endowment that keeps its self-management ability intact over the long term. For the endowment's ecosystems, this means retaining their resilience and robustness, and their ability to self-adapt.

---

[17] Felitti, V. J. 2009. Adverse childhood experiences and adult health. Academic Pediatrics, vol. 9, pp. 131–132; and Felitti, V. J., Anda, R. F., Nordenberg, D., Williamson, D. F., Spitz, A. M. et al. 1998. Relationship of childhood abuse and household dysfunction to many of the leading causes of death in adults. The adverse childhood experiences (ACE) study. American Journal of Preventative Medicine, vol. 14, no. 4, pp. 245–258.

It may seem that the significance of the environmental conditions for viability have been exaggerated. After all, our adaptable culture and our mastery of technology has allowed humans to live in a wider variety of environments than one would expect. More personally, consider the cornucopia of goods that are available to us. But this view neglects that, for a complex culture, such success comes with increased responsibilities and the need to face inescapable resource limits.

Certainly, compared to a simple culture, a complex culture's greater access to technology gives it a much greater capacity to modify its environmental endowment and make more resources available. However, because of its greater population and the larger per person cost of its reaffirmation and authority maintenance efforts, a complex culture, especially a more complex one, requires significantly more (absolute and per capita) resources than a simple culture to function. And, in the process of acquiring and using those resources, a complex culture can more easily disrupt its supporting environment. Think of direct action (e.g., deforestation, soil erosion), pollution and climate warming. At some point, these disruptions burden a complex culture with more responsibilities than its members can appropriately discharge, which makes it extremely difficult for the culture to become or remain viable. Thus, if the members of any culture want to favour viability, then they must take the environmental constraints on viability seriously. For example, they must live more in tune with the long-term reality of how resources are supplied and the consequences of using and discarding them.

Regardless of whether we are members of the public or the elite, we are, individually, our culture's decision makers. It is we who have to make and implement the decisions needed to achieve viability. Our striving toward it begins when we accept that the resimplification of our complex culture is highly likely, and realize that we have the capacity to avoid the worst of a forced resimplification. We are then more likely to realistically evaluate the option of voluntary resimplification. And we are more likely to proactively make the (material, intellectual and emotional) changes needed for our decisions to favour us living within viability's boundary (figure 3).

In essence, we are each required to both find a way to live within viability's environmental and human constraints, and make the mental and material changes needed to do so. We are obliged to change our personal world view. This includes changing the way we think about our world and ourselves, and how we relate to our environment and one another. In particular, the changes must reduce the divergences in our internal reality so that our personal world view more clearly reflects the external reality. Similarly, we must reduce our responsibilities, and favour viability when we fulfill those we are stuck with.

Making these decisions and changes is a complex and ongoing task. It is like the search for happiness: a work in progress, a state of becoming. It is a quest rather than a set of simple, deterministic, material solutions. And our efforts can easily be sidetracked or co-opted: for example, if we decide to "reduce" our responsibilities by simply ignoring them.

Fortunately, if most of us successfully devote the time and effort needed to change our personal world view and lifestyle to favour viability, then we will also, collectively, make similar changes to our culture's world view. It too will then preferentially guide us to live and make decisions in ways that satisfy viability's conditions (figures 2 and 3). In doing so, we will form a member-culture feedback that favours viability. But, as hinted at, we will encounter challenges in making these changes. They are discussed here by dividing them into four overlapping source groups.

Challenges Originating from Individuals

Some of the challenges we will face when reaching for viability will originate from our innate characteristics, a few of which were discussed in the subsection "The Decision Makers: Individuals" in chapter 3. They include our flashlight view of the world, our limited appreciation of scale (of time, space and intensity), our decision-making foibles and our divergences.

Although we can't, as some suggest, "fix" or get rid of these innate characteristics, we can recognize and accommodate them, and their products, in ways that will help us to make decisions that favour viability. But doing so has its challenges. These are illustrated by the characteristics we must deal with if we are to recognize, and accommodate or reduce, those of our divergences that critically affect our efforts to favour viability. For example, we innately prefer to treat the human-environment system as a simple deterministic system and think about it in a black and white manner. We also prefer to personalize and assimilate information that is human-centred or we feel is acceptable rather than test its truth or search for information that is more reliable. Overall, it is easier for us to follow our urges and our culture's guidance than take the time and make the effort to broaden our understanding of ourselves and our surroundings in order to address our critical divergences.

Consider too that it takes time to bridge the gap between theoretically accepting that an issue, such as a divergence, exists and actually engaging in practical action to deal with it. Thus, if we want to address the issue in a timely manner, then we must start making the needed changes well in advance of wanting to see the results. But we also have a strong, culturally supported innate inertia to change, so we prefer to be reactive not proactive. These types of challenges are difficult to deal with. Indeed, as a result, we are likely to continue for as long as possible making decisions that favour us experiencing a forced resimplification. This outcome is made more likely when the members (elite and public) succumb to the temptation of adopting a personal world view that favours those aspects of their culture's world view that support their personal preferences in the short term, but not viability in the long term.

Fortunately, if enough of a culture's members (a critical mass) do make most of the changes needed for them to favour viability, then they can, collectively, change their culture's world view so that its guidance does the same. A challenge to reaching that critical mass is the diversity of opinions among the members about the need to change and how to resolve those differences, as can be seen in

Western countries today. Consider too that the members also face Catch-22 challenges: feedbacks that restrain their efforts. For example, the members can change their culture's world view because they are both the carriers of their culture and its decision makers. But, their efforts to make and implement those decisions are influenced by their culture's world view. This completes a member-culture feedback that favours maintaining the status quo.

There are other challenges the members face in their efforts to ensure their culture favours viability. Some cultural guidance does favour viability, which the members must recognize and then refrain from (deliberately or inadvertently) corrupting or ignoring. We fail to do so when we, for example, form culture-disrupting, antagonistic social cliques, gangs and equivalent businesses, and when we individually, especially if we are from the elite, fall into the trap of narrow, self-serving visions (e.g., groupthink) and succumb to the temptations of more riches and power. Examples are Conrad Black for Hollinger, Ken Lay for Enron and the leaders of the many large financial institutions who contributed to the global financial crisis of 2008.

Our relationship with our culture can lead our efforts astray in other ways. For example, when a culture's guidance is given in the form of a symbol, then much of a member's response to it takes place in their unconscious. Think of advertising and brands. This allows us to respond more rapidly than if we used conscious thought, but it also reduces our opportunity to consciously review our response. This increases the possibility of us responding inappropriately or unquestioningly to the display of cultural symbols. For example, a leader can use the power of a flag to gain the members' support to be the aggressor in an unnecessary war that is counter to viability. Think too of words as symbols and the rise of populism.

We burden ourselves with a different set of challenges when we are distressed. When enough of a culture's members are distressed, they can more easily respond to issues in ways that are at odds with viability. For example, a common response to the loss of a shared cultural value or belief, such as our faith in our culture's institutions or myths, is to adopt a more mystical personal world view. If enough of us adopt this view, then it can become part of our culture's world view. Such mystical views need not be benign. An example is the rise in concern about witches that occurred in Europe between the late medieval era (mid-1300s) and the start of the Enlightenment (late 1600s). It was one product of the profound religious, health and structural changes underway during that period.[18] Think too of contemporary unsubstantiated conspiracy theories and ideas of colonizing planets around other stars.

Challenges Originating from Culture

Some of the challenges we will face in our efforts to favour viability will originate from our cultures. Consider that, to be functional, a culture must satisfy a suite of basic requirements and appropriately resolve any conflicts that might arise when trying to simultaneously satisfy them. The following are examples of those requirements and the conflicts that arise between them: a tolerance of diversity and the promotion of a strong sense of identity; cultural flexibility and faith in the culture's existing world view; a focus on the group and on the individual; the need for surveillance and privacy; and enabling individuals to reach for their personal aspirations while also protecting individuals and the environment against exploitation. In addition, a culture's world view must refrain from forcing good people to do bad things. It is a challenge for a culture (i.e., its members) to simultaneously satisfy all of these requirements, let alone consider viability's other conditions.

A culture is also a source of challenges because its world view influences its members' decision making. As mentioned above, the member-culture feedback ensures that a culture's world view, especially its core, displays inertia to change. For example, the notion of progress has been a prominent feature of Western culture since at

[18] Roper, L. 2004. Witch crazes: Terror and fantasy in Baroque Germany. Yale University Press, New Haven.

least the mid-1800s. Consider too that Western culture's guidance about our relationship with the environment and technology makes it easier for us to falsely believe that the simple application of personal effort, combined with our ability to innovate and create technology, can largely solve our human and environmental problems. This is illustrated by its guidance that favours us solving our environment-related issues in ways that use technology and maintain economic growth. We can thus believe, for example, that the disruption of the environment resulting from the extraction and use of resources could be significantly reduced through just a technology-facilitated increase in their efficiency of use. This increase will, so we trust, automatically result in an absolute reduction in resource use, and thus a decline in the associated disruption of the environment.

But our belief leaves out the question of what actually happens to the savings from increasing efficiency. In general, spurred on by our culture's belief in economic growth, they are used to maintain and promote resource use. Think consumerism and the virtuous economic cycle. For example, the energy saving from replacing incandescent light bulbs with more energy efficient LED lights is being consumed by us installing more or brighter lights on more things. The savings from increasing efficiency are thus neither, literally, left in the ground nor, figuratively, put back into it, so environmental disruption continues. Our belief in guidance that is incomplete or largely false in practice hinders us from adjusting our personal and thus our culture's world view so that it guides us to make the decisions that will actually result in an absolute reduction in both our use of resources and our disruption of the environment.

Overall, a culture's inertia to change, in combination with the influence it exerts on its members' decision making, ensures that the members will face a challenge when they try to change their personal and their culture's world view so that it guides them to favour viability. Perhaps the most serious of these culture-sourced challenges arises when a culture guides its members to deal with issues in ways that not only conflict with viability but direct them to ignore viability's boundary. For example, Western culture's notion of progress, a preference for technological solutions to issues and an insistence on economic growth support a feedback loop that favours an ongoing, unsustainable increase in Western culture's complexity and resource use.

Challenges Originating from the Environment

Challenges to making viability-favouring decisions can also originate from the environment. Their presence and nature are revealed when we consider how some of our laudable efforts to do the "right thing" failed. Examples are the introduction of carp into the southern United States to clean fish-rearing ponds, and starlings into Australia to control insects, both of which became invasive pests.

Consider too the more recent efforts to maintain wild pollinator populations primarily by managing the land, such as nurturing exotic flowering plants to sustain them, rather than protecting their habitat. These types of actions may succeed in the short term, but in the long term they simply shift, transform or worsen the pollinator problem. They also burden us with additional (short- and long-term) environmental responsibilities (e.g., dealing with invasive plants) that we are unlikely to appropriately fulfill.

Finally, consider that working conscientiously toward lowering a country's infant mortality rate is a laudable goal, except that it will also likely increase the country's population size. That will result in an increasing demand for food and water, which will, in time, exceed the capacity of the country's environmental endowment to support its now larger population. This will have a negative effect on the health of its members, including the new infants, which runs counter to the aim of lowering child mortality. It is a challenge for us to recognize that our goal of lowering infant mortality for the long term can only be successful if we also simultaneously work toward lowering the long-term fertility rate.

It could be argued that more effort could be put into increasing the availability of food and water. But, similarly, the presence of fundamental environmental constraints on these resources ensures that those efforts too can only delay undesirable impacts on infants, including a rise in mortality rates. We could decide to believe that an increase in population size is always a good thing or that higher infant survival will itself automatically lower fertility rates, but that is gambling.

These examples point to the basic kinds of interconnected, environment-sourced challenges we will face when we try to favour viability. They include the following. Our actual relationship with the environment is poorly reflected in our culturally mediated personal relationship with it, which hinders our ability to make decisions that are truly environmentally friendly. Our interactions with our environment invariably create environmental responsibilities that we should discharge in an environmentally friendly manner; but doing so is a challenge that can be exacerbated by our culturally preferred relationship with our environment and feedbacks. Our short-term interactions with our environment can result in long-term cumulative adverse effects, unless we gain the needed vision/awareness and effective (personal and cultural) processes to acknowledge those impacts, and then limit our interactions to environmentally acceptable levels. Think of the increase in population size from a focus on just reducing infant mortality. And finally, we need to widely coordinate our activities and undertake oversight checks to ensure that they favour viability. If these types of challenges are not met, then the results of our environment-affecting activities are unlikely to satisfy viability's conditions.

The overall nature and significance of the environment-sourced challenges we face, and the difficulties of dealing with them in a way that favours viability, is illustrated by the meaning Westerners usually intend when we use the words *adapt* and *sustainable*. In the context of our current circumstance, Westerners invariably use *adapt* to mean responding to our changing environment in ways that maintain the essence of our current lifestyles and beliefs. It rarely means the alternative: changing our relationship with the environment so that our interactions with it and our lifestyle more closely reflect the new reality and viability's constraints. Changing our view of *adapt* is a challenge.

The term *sustainable*, which expresses the idea of maintaining something over the long term, has become a reference standard or goal that is commonly used when discussing what to do about our current circumstance. It came into prominence as a result of the 1987 Brundtland report on sustainable development, which proposed that focusing on it would resolve both our social and environmental issues. However, the report's wording allowed the term to be defined and objectives to be selected that don't fully account for the complex, collective nature of the constraints on how we can live. As a result, *sustainable* has taken on meanings that are contrary to its viability-favouring roots (figure 3).

This can be seen in the recent United Nations report on 169 goals to achieve sustainable development. Although many of the individual goals are laudable, reaching for them will only favour viability if we ensure that they don't conflict with one another and that our efforts to reach for them don't work at cross purposes: our goals and efforts need to be coordinated, subject to oversight and merged to ensure that the greater goal will be achieved. Think infant mortality.

This is not meaningfully the case. For example, goal 8 aims for "sustained, inclusive and sustainable economic growth" (i.e., continuous economic growth).[19] This goal reflects the primacy of achieving economic growth in industrial cultures and allows (neoliberal) economic assumptions to be used as a primary filter when trying to decide if efforts to reach it are succeeding. We already know that a focus on this goal limits, co-opts, misdirects and even subverts our efforts to protect the functioning of our environment. Thus, striving for goal 8 renders efforts to reach the goals that protect the environment and satisfy some aspects of human welfare ineffectual, if not meaningless. Consider too that

[19] United Nations. 1987. Towards sustainable development: Our common future: Report of the World Commission on Environment and Development. Published as Annex C to the General Assembly document A/42/427. Chapter 2, pp. 1–21. http://www.un-documents.net/our-common-future.pdf; and United Nations. 2015. Transforming our world: The 2030 agenda for sustainable development. United Nations General Assembly, 21 October, 2015, A/RES/70/1 7th session, agenda items 15 and 116, pp. 1–35.

reaching for continuous growth of any kind is incompatible with the reality of living on a finite planet. In this context, the term *sustainable economic growth*, as used in goal 8, is absurd. It is like *sustainable mining*, which is nonsensical because the process deals with resources that are non-renewable on a human-relevant time scale (finite in a practical sense).

We would do well to remember that becoming *sustainable* and success at *sustainable development,* as the terms are currently used, do not guarantee that the environment system, our cultural systems or the human-environment system will become functional or remain so over the long term. *Sustainable* is thus not a synonym for *viable.*

If we wish to live within viability's environmental constraints, then we are obliged to face these environmentally sourced challenges, which is itself a challenge.

Challenges Originating from Complexity

As hinted at above, the complexity of the human-environment system and its subsystems will itself present us with challenges. This is unsurprising because we tend to find complexity and its expression in systems hard to grasp, confusing, baffling, surprising, intractable and even frightening.

How did you relate to the earlier description of a complex system's functioning? It was described as the expression of the system's many variables interacting with one another over a wide range of time, space and intensity in a variety of ways that include feedback. Change was described as the system responding to a flexible hierarchy of ongoing (internally generated and externally imposed) influences on its functioning. These features of complexity are difficult enough to grasp, especially when compared to our preference for simple, deterministic cause and effect. But you then have to also consider that the description of a system's changing is just a different view of its functioning, and that processes are just as important as causes and effects.

If we are to successfully favour viability, then we are faced with the challenge of personalizing and assimilating the features of complexity and their expression in the human-environment system. This includes accepting that our difficulty thinking about complexity and complex systems is a result of more than complexity itself. It is also a reflection of our human perceptual and decision-making foibles, and the influence of our culture's world view on our lives and our choices. It also includes accepting that, at some level, we are each more than disinterested observers or managers of the human-environment system: we are active participants, with feelings, who are embedded in it.

We also face the challenge of including the characteristics of complexity in our decisions. Consider the challenge of identifying the inescapable limits and constraints on the order and predictability of the complex human-environment system and taking them into account in our decisions. For example, our predictions about an aspect of the system will always contain inescapable uncertainty, and we should therefore temper our decisions and actions accordingly. Think precautionary principle.

If we wish to spend our time searching for and implementing viability-favouring solutions to a complex issue rather than arguing about concocted or perceptual differences, then it is essential that we face the challenge of dealing with complexity and its expression.

***

In summary, chapter 5 discussed how reaching for the goal of voluntary resimplification can help us work toward the alternative future of viability, and the challenges of doing so. In essence, we are our culture's decision makers. So it is up to us (public and elite) to make the decisions that will change ourselves and our culture so that we more fully satisfy viability's human and environmental conditions. They must be satisfied both individually and collectively in their form of a viability boundary (figures 1, 2 and 3).

The challenges we face arise from the characteristics of the human-environment system. For example, we are challenged to make, rationalize and implement decisions that have considered our characteristics as individuals, the features of our culture's world view, the nature of our environmental endowment and the complexity of the dynamic interactions among them. We are also challenged to coordinate our activities and routinely undertake oversight checks.

Many of the human-related challenges we face are not correctible (removable) human imperfections but part of who we are. We can't change these features, but we can accommodate them. In particular, although we can't change our thinking processes, we can change how we direct our thinking. We can acknowledge our human foibles, such as our decision-making characteristics, how strongly we are tied to the routines of our day-to-day lives and how deeply we are embedded in our culture. We can accept how mentally distant we are from our environment, our dependence on it and its complexity. When all of this becomes real to us, personally, we are more likely to accommodate it in our decisions in a way that favours viability.

If we are personally prepared to expend the time and effort over the long term, then we can gain the wisdom needed to meet the challenges of making viability-favouring decisions, which is the subject of the next chapter.

Chapter 6

ACHIEVING VIABILITY

If we wish to live in a viable culture, then we, our culture's decision makers, must transform ourselves so that our decisions routinely favour viability (figure 3). This chapter discusses ways to achieve that objective. But first it provides a summary reminder of the task ahead: the changes we must complete, the aids available and the challenges we will encounter along the way.

The Task

We are faced with recognizing and accepting our human foibles and their limitations for what they are. And, because they are an influence on our decisions that can't be removed, we are obliged to compensate for them in ways that are consistent with the external reality and viability's conditions. We are obliged to come to terms with the complex functioning of the human-environment system. In practical terms, our decision making will have to accommodate the system's features, such as a diffuse and enigmatic boundary, somewhat flexible constraints, feedbacks, a wide range in scales (of time, space and intensity), uncertainty and grey choices, and do so in ways that favour viability. We will also have to balance conflicting interests and constraints, so we will have to be open to compromises and trade-offs. But, because viability is the reference and goal, there can be no compromise on living within viability's conditions. Because of the inertia and feedbacks in the system, our efforts to satisfy viability's conditions will be ongoing. Fortunately, given time and with effort, we can adjust ourselves and thus our culture so that the feedbacks favour viability.

These changes are easier to make if we take advantage of the available mental and material aids (figures 1, 2 and 3). In particular, the descriptive model of the complex human-environment system and the viability flower help us make sense of our circumstance, identify issues and possible solutions, and evaluate them. They also help us to appreciate the benefits and limitations of predictions and mathematical models. The aids do so because they prompt us to pay attention to the system's complexity: its dynamics, feedbacks and a wide range of time, space and intensity. They remind us to consider context, constraints and both material and mental aspects. We are also prompted to look in the mirror: to accept our foibles and pay attention to our rationalizations, responsibilities and divergences. Overall, the aids help us to both make and question our decisions in ways that favour viability.

The scientific method can help too, because it provides us with the most widely available, reliable (independently reproducible) information about the human-environment system. But it is insufficient on its own to ensure that we will meet viability's conditions. Striving for viability requires us to be more than a skeptical, disinterested observer of the system. At some point, we must include ourselves in it as participants who have an internal reality, human foibles, a spiritual-emotional dimension and a vested interest in the future. The model of the human-environment system and the viability flower provide that reminder. So can religion.

We can also use the reliable knowledge we gather to develop and use rules of thumb that favour viability. We can then apply them, within their limitations, as decision-making guides. Similarly, we can choose objectives that reflect viability's human and environmental conditions/constraints and work toward them. Here are five examples of such environmental objectives: (1) accept the environment's supremacy; (2) maintain its functioning; (3) champion its self-management; (4) conserve what we have; and (5) reduce our total (absolute) resource demand. Objectives can even be bundled together into subgroups to emphasize their interdependence and interactions. For example, the above environmental objectives can be grouped into a "five-objective unit." By striving toward these objectives, singly but especially when bundled, we provide ourselves with a feedback-assisted incentive that can, along with coordination and oversight, help us reach for viability. They can also, collectively, become a long-term rallying cry: Reach for viability! Accept voluntary resimplification!

If our efforts to strive for viability are to be effective, then we have to cross the gap between theoretically accepting that an issue needs our attention and undertaking the personal action needed to appropriately deal with it. Crossing this gap is difficult at the best of times, as anyone who has tried to break a habit, lose weight or reshape their corporate plan knows. It is particularly difficult when striving toward viability because doing so usually requires us to reduce something that is normal for ourselves or our culture, such as our level of resource use. Think lifestyle. We help ourselves to cross the gap by creating a stepping stone. It can take many forms, such as adjusting our beliefs, setting a series of intermediate goals or finding people to help. The descriptive model's comprehensive picture of our circumstance is also useful here. It helps us to create the safe mental place we need if we are to tackle the challenges we face.

We (public and elite) also help ourselves to cross the gap, and stay on the other side, by asking ourselves personal questions such as, what does achieving viability require me to do more or less of? And does this decision (act, purchase) favour viability? More specifically, will this decision result in me walking more softly on the land? Will it reduce my responsibilities? And, if so, will the method of reduction favour viability? Is the increase in convenience or freedom of choice I will receive from this decision really worth the associated increase in my environmental, social and personal responsibilities?

In the specific case of resources we can ask, How much stuff is *actually* enough to ensure my material and mental satisfaction? And its corollary, What is an appropriate level of self-rationing? We can also ask of a purchase or action whether there is a more viability-friendly (material or mental) alternative/replacement. Then there is the question of whether the savings from a corporate or personal proposed increase in efficiency or reduction in waste will actually result in an absolute reduction in resource demand or environmental disruption. Basically, will the savings be, literally, left in the ground or, figuratively, put back into the ground? Asking these questions is just as important as arriving at an answer because we prompt ourselves to expand our thinking.

It should now be clear that there is no one, simple action, such as a focus on renewable energy, that we and our culture can follow to become viable. We are stuck with continually making decisions over the long term about a wide variety of dynamic issues in different, but always complex, contexts, while constantly considering viability. Success requires us, in essence, to gain wisdom and routinely apply it.

Wisdom is more than the awareness of information (knowledge), the casual, unquestioned use of knowledge or the skilled use of knowledge (intelligence). It is the ability to know when you don't have adequate information, to collect the needed knowledge and to know when and how to use it appropriately. Wisdom is the skill to use the knowledge you have as an aid to making judicious assessments of the many factors affecting an issue when it is seen within its larger context, and to then decide between, or delicately balance, the accessible options. Wisdom is not how much you know, but how you use what little you know.

With this in mind here are three methods that can help us personally gain the wisdom needed to address the above kinds of tasks and the associated challenges in a way that favours viability: (1) increase our knowing; (2) engage in nuanced thinking and decision making; and (3) tweak our personal world view. Think of these methods as specialized parts of the rationalization process in the human-environment system that links the belief and resource loops together (figure 2). When used with a guide to viability and the environment's functioning, they can move us toward viability (figures 1 and 3).

Increase Our Knowing

We can more easily make and implement viability-favouring decisions if the information available to us and how we use it better reflects the external reality. The scientific method of gathering and checking information, whether applied formally or informally, provides us with that information. But, as previously mentioned, if it is to help us reach for viability, then reliable knowledge can't remain detached from us. It must cross the gap. It must become a part of ourselves and be used in our decision making: we must use it to increase our knowing.

We start to increase our knowing when we attach considered personal meaning or relevance to the most reliable knowledge available to us about the external reality: when we personalize reliable knowledge. This can be more difficult than we expect because the knowledge we discover can repel, overwhelm or dishearten us. Think complexity. Therefore, part of personalizing knowledge is to come to emotional terms with it. Once the reliable knowledge we have gathered has been personalized, we can merge it into our personal world view to become part of who we are: we can assimilate it into our internal reality. (Although mentioned separately, personalizing and assimilating can occur together.)

Reliable knowledge that has been personalized and assimilated changes our internal reality so that it more closely matches both the external reality and our actual relationship with it: it reduces our critical divergences. If we then use that information in our decision making in ways that change our lives, we will have demonstrated that we have increased our knowing. Knowledge has crossed the gap. For example, after feeling the heat of flames, we know that fire burns us. Deciding not to touch them shows that our decisions now better represent the environment, and us, for what they actually are. We have taken this reliable piece of knowledge to heart: we have increased our knowing.

It is convenient to talk about increasing two types of knowing: knowing about ourselves, our self-knowing; and knowing about our environment, our environmental knowing. Both types and how they are interconnected were illustrated by the example of fire. Increasing both types of knowing is a key aspect of gaining wisdom and making wise decisions.

In overview, by personalizing and assimilating reliable knowledge about the characteristics of humans and the environment, and then using it to make decisions, we demonstrate that we have increased our self-knowing and our environmental knowing: we have increased our wisdom. We are then more likely to accept viability's conditions as goals, and make the decisions and changes needed to satisfy them. In particular, it is easier for us to adopt a world view and lifestyle that better matches the external reality.

Practise Nuanced Thinking and Decision Making

Although we can't change the basic processes by which we think and make decisions, we can change how we direct them. We can think and make decisions in a way that better reflects the complexity of how we, our culture and our environment function: we can adopt a more nuanced manner of thinking and decision making (nuanced thinking). What does this require us to do?

I often puzzle over how I can hold multiple, often contradictory opinions about our neighbouring forest. I can think that it should be exploited for wood and recreation, nurtured for the food and water it supplies, or respected because its inhabitants have an unquestioned right to exist. I also note that I can switch between my opinions with little recognition of the contradictions or conflicts among them.

A resolution to my puzzle starts to appear with the realization that each of my opinions about our forest arise from a different perspective of a single dynamic, complex forest, in which I am a participant. Thus, when making decisions about our forest from one perspective, I should fully consider the other possible perspectives I could take of it, and the links between them. This multiperspective view of our forest provides the clarity of simplicity, while also adequately representing its complexity and the limitations of my understanding about it. But this view contrasts with my culturally supported, innate human tendency to perceive and think about our forest by focusing on one "independent" aspect or perspective at a time.

I realize that to engage in a more nuanced manner of thinking means adopting a broad (multiscale), constructivist, comprehensive (multiperspective), flexible (open-to-change) manner of thinking and decision making. Although nuanced thinking uses the techniques of scientific thinking, such as being skeptical (including being self-skeptical), it is more than scientific thinking. It is directed at more than acquiring factual knowledge. It is a striving to gain understanding about an event/topic and make wise decisions about it. Nuanced thinking thus requires that we each treat ourselves as a mix of an objective (disinterested) observer, an emotional observer and, at some level, a participant. We are obliged to consider when to apply each during our decision making.

In order to engage in nuanced thinking about a particular topic, I must consciously take into account complexity and context. I must remember and consider that I am part of a dynamic, complex human-environment system in which I am simultaneously an individual, a member of a culture and a member of my culture's environmental endowment. Because I participate in the system, I am both part of its functioning and I am constrained by it (figures 1, 2 and 3).

Similarly, I must accept that the topic I am considering is itself a part of the human-environment system, so despite my desire to, I can't treat it in a simple black and white manner. It, and the context in which it resides, is also complex, dynamic and grey. And I am, in some way, connected to it.

Thus my contemplation of the topic has to consider many context-dependent views of it. In doing so, I must remember that contradicting observations and opinions about the topic can arise from differences in perspective, as well as from errors in knowledge or thinking. I am also obliged to recognize and accommodate in my thinking the existence of my internal reality's divergences, my decision-making foibles, the influence of my culture's world view (including its divergences) and the environment's complex functioning. And, to favour viability, I must always take the conditions for viability into account: they are a primary reference (figure 3).

Success is more likely to be achieved if I also work at increasing my (self- and environmental) knowing. Fortunately, the two methods are complementary parts of gaining wisdom. Consider the question of whether our unique human characteristics make our species superior to all others. Or are we just another species whose implied ordinariness means that we are worthless? If we can both increase our self-knowing about our need to feel special and practise nuanced thinking (which moderates our tendency to focus strongly, especially on ourselves), then we can arrive at and accept the following answer.

Each of us is a member of a human community where our role as an individual is special. But each of us is also a member of the human species, which is part of the human-environment system. In this system, we are each personally subject to the same basic rules as every other living thing. In that context, we are ordinary. We are thus both special and ordinary at the same time. Which view is appropriate depends on the context or perspective we choose. Regardless of our view, for it to be appropriate it must favour viability.

Tweak Our Personal World View

Our efforts to practise more nuanced thinking and increase our knowing in order to favour viability changes our personal world view. But our personal world view influences our efforts, thus forming a feedback loop. Similarly, any change to our personal world view prompts our culture's world view to change. But our culture's world view influences how our personal world view will change, which completes another feedback loop (figure 2). Although these interconnected feedbacks help to stabilize our personal and our culture's world view, they can also constrain our effort to better practise nuanced thinking, increase our knowing and favour viability. Our efforts would be helped if we could directly tweak our personal world view.

Surprisingly, there are opportunities to do so. For example, a complex system displays random and chaotic, as well as deterministic, features. Yet a member of at least Western culture tends to discuss and make decisions about complex systems using a deterministic vocabulary. By simply adjusting our vocabulary so that the words we use more accurately reflect the system's actual features, we directly tweak our personal world view in a way that boosts our efforts to practise more nuanced thinking, increase our knowing and thus favour viability.

Such an adjusted vocabulary has been used in this book. For example, when describing human behaviour, our decision making and environmental processes, *tend* was often used instead of *is* or *do*. Here is an illustration of why it was used when referring to people. There are many ways an individual can respond to something. However, when the responses to it from a number of people are considered collectively, then the group often displays a preference for just a few responses. Thus a member picked at random from the group is more likely to use those preferred responses: individually we tend to prefer one of those responses. For example, when the members of a group are asked to rate their driving skill, the majority, but not all, will rate themselves as above average: we individually tend to think we are an above average driver. An example of the use of *tend* in this book can be found by rereading the third sentence in the above paragraph.

Here are a few other tweaks to consider. Instead of referring to sustainable development (or sustainability) as a long-term general goal for our future, refer to viability. And take care to make the distinction between *sustainable* and *viable* quite clear. Similarly, when talking about evaluating a proposed environment-affecting action, refer to the need for a responsibility-benefit analysis rather than a cost-benefit analysis. And the word *progress* should not be used as a synonym for the goal of *growth* or *development* of any kind, especially not economic.

A more personal example suggests that instead of using *right* and *wrong* when evaluating our personal (or other people's) thoughts and decisions about a human-environment topic, we can use *appropriate* and *inappropriate*. Their inherent broader meaning reminds us to consider a topic within the wider, more complex context in which it usually exists: the dynamic, complex human-environment system. Using these two words thus prompts us to think and make decisions about the topic in a more nuanced, rather than a black and white (simple, deterministic), manner.

The use of *appropriate* and *inappropriate* will also help us to apply our judgement about a topic in a more nuanced manner. For example, the making and then marketing of an aquarium fish that glows in the dark (the GloFish) for our amusement is certainly environmentally inappropriate: the modification radically disrupts a species' lifeway and increases our environmental responsibilities. It should also be morally unacceptable in cultures that see themselves, whether for secular or religious reasons, as environmental trustees.

There is a constraint on the use of *appropriate* and *inappropriate*. Our knowledge and appreciation of a complex topic is always incomplete to some degree. There will thus be situations where our level of uncertainty means that we are unable, despite our desire, to meaningfully assign *appropriate* or *inappropriate* to our thinking and decisions about a topic. We could just add a conditional term to our assignments, such as *appears*, *seems* or an equivalent qualifier. But here the discussion turns to a broader generic replacement word: *incomplete*.

By applying *incomplete,* we can prompt ourselves to search for more reliable information. But it takes effort to use it. Consider that we often find the notion or sense of incompleteness unsettling. Think of the emotion behind the phrase "seeking closure." This need for a sense of completeness is not a weakness. It is an inescapable aspect of how we innately and culturally deal with the reality of living in what is, to us, a sometimes unsettling, unpredictable, complex world. Think of how we use phrases such as "an act of God," "things will be fine" and "an unforeseeable accident." Think too of how rituals finalize a circumstance.

Remember too that it is also normal for our brain's unconscious processing to automatically fill in any blanks in a perception or thought before we become conscious of it. Similarly, when a conscious thought or decision feels unsettlingly incomplete to us, we intuitively call first on our unconscious thinking and our culturally influenced personal world view to help us deal with it. We usually find (or more accurately, we are provided by our brain with) a solution that is appealing and decisively resolves the issue. By accepting it, we gain both a logical and an emotional sense of completeness. But, if the satisfying solution hasn't been checked for reliability, then it can allow us to mistakenly believe that the use of *incomplete* is unnecessary. Thus, on one hand, the term *incomplete* can help us to more reliably reflect on our limited understanding of a topic, but it can also leave us emotionally hanging and labelled as waffly. On the other hand, relying on an unsubstantiated belief can provide us with the emotions we desire, but it may not help us to reliably reflect on our limited understanding of a topic.

Fortunately, this issue can be resolved by making another adjustment to our personal world view. We can hold the view that forming and holding beliefs is part of being human and that the beliefs embedded in our personal world view help us to deal with unsettling experiences and difficult problems. But we can also accept that our beliefs do not necessarily represent the external reality: we leave the door open to adjusting our beliefs and applying conditionals to them. In this way, we gain the benefit from our beliefs but avoid being trapped by our blind faith in them.[20] It also becomes easier for us to practise more nuanced thinking and increase our knowing, which completes a feedback loop whose features can help us deal with uncertainty and the unknown, and do so in a way that favours viability.

By making external-reality-compliant adjustments to our personal world view at the appropriate time, we can help our efforts to engage in nuanced thinking and increase our knowing, and thus favour

[20] Moshman, D. 1998. Cognitive development beyond childhood. In Kuhn, D. and Siegler, R. S. (editors). Handbook of child psychology, 5th ed. Vol. 2. Cognition, perception and language. Wiley, New York: pp. 947–978.

viability. We can help ourselves to make additional appropriate tweaks by asking ourselves questions such as, what modifications would help me live a skeptical, spiritual life?[21] Which adjustments would help me exhibit humble scientific thinking?[22] If we focus on viability, then we could ask: What tweak would help us to not only resolve the conundrum of needing to both revere and transform the environment, but do so in a way that favours viability? In particular, what tweaks would help me become a trustee for the environment (protecting its functioning) rather than a manager or steward (favouring its exploitation)?

***

In summary, chapter 6 described three methods to help us (public and elite) individually favour viability: increase our knowing, tweak our personal world view, and practise nuanced thinking and decision making. It also described how our efforts to apply one helps us to apply the others: applying them is mutually reinforcing.

By applying the three methods, we help ourselves to overcome the challenges and make the personal changes needed to reach for viability: they help us transform ourselves to favour viability. In particular, they help us come to terms with our human characteristics, and appreciate both complexity and how the complex human-environment system functions/changes. They also make it easier for us to apply the aids, such as the model of the human-environment system, that help us to understand issues and find solutions that favour viability.

The three methods also help us to make the decisions that will bridge the gap between our theoretical acceptance of an issue and our practical implementing of an appropriate solution to deal with it. And, once we are on a viability-favouring track, they make it easier

---

[21] Raymo, C. 1998. Skeptics and true believers: The exhilarating connection between science and religion. Doubleday Canada, Toronto.

[22] Ziman, J. 1978. Reliable knowledge: An exploration of the grounds for belief in science. Cambridge University Press, Cambridge.

for us to stay on it during our day-to-day living. Here are some simplified examples of such personal (public and elite) decisions: adopting diets, farming practices and recreational activities that are less disruptive of the environment; reducing our resource use and ensuring that the savings are "put back into the ground"; paying a living wage; and reducing the income gap (e.g., by not striving to acquire excessive wealth). By these means, our personal efforts can both transform our culture's world view and establish a feedback that will, collectively, guide us to favour viability over the long term.

Chapter 7

THE OUTLOOK

Collectively, the previous chapters form a framework that helps us to gain a comprehensive and coherent appreciation for our current circumstance, and consider our possible futures. It also helps us to evaluate possible solutions to issues, select the most appropriate ones and implement them. Using it provides the following summary outlook for our future and an evaluation of how we could most appropriately respond to it.

The inertia displayed by some of the trends leading to resimplification is sufficiently large that the consequences can't be completely stopped. Other features of resimplification will arrive because of a lack of either will or wisdom to appropriately deal with them, such as concluding that an appropriate way to reduce our responsibilities is to abandon/ignore them. Some features will arrive because of the counterproductive co-opting, misrepresentation and corruption, especially by the elite, of the efforts to slow their arrival. Consider greenwashing, where false claims about going green are presented as true. Familiar examples are asserting that efficiency will, on its own, automatically reduce resource use (it won't) and that carbon offsets are a green compensation for emissions (usually an illusion). So is the statement that shipping liquid natural gas overseas will automatically reduce coal use there (it's not written in the contracts).

Resimplification will also arrive because our responses are ineffective, in particular, the elite failing to accept and discharge their extra responsibilities in a way that favours viability (e.g., by focusing too strongly on self-interest and ideology). More general ineffective responses are those that place too strong an emphasis on the material (e.g., a focus on hoarding supplies), anything but the material (e.g., we just have to work harder or be better people), or a single issue (e.g., green energy). Responses are also ineffective if they rely on fear-mongering, false hopes and reaching for unrealistic deadlines. In essence, beyond inertia, forced resimplification will arrive due to the triumph of actions and beliefs that ignore or unrealistically consider viability's constraints or the functioning of the human-environment system.

Striving for viability is certainly the most difficult to implement of our possible responses to our current circumstance. After all, even with the help of the related aids, it still takes our precious time and effort to personally apply the three methods and to become familiar with both complexity and the complex human-environment system. Fortunately, by firmly persisting in doing what we can, we will be gaining the wisdom needed to change ourselves and our culture in ways that favour us living within viability's human and environmental constraints: within viability's boundary. Thus, by starting now, we are collectively more likely to at least reduce the severity of the impacts that a forced resimplification will have on our lives and the environment. We will favour voluntary resimplification.

In addition, over the long term, our efforts will have primed (preconditioned) future generations to make decisions and live in ways that favour viability (figure 3). And we will also have facilitated the creation of a feedback that will favour them continuing to reach for viability over the long term (figure 2). This will allow us to retain a sense of pride in ourselves, our humanity and our culture. And our children will be happy that we at least tried, and that we preconditioned them to continue striving to meet the conditions for viability.

Further justification for this conclusion, along with the details that will help your personal efforts to favour viability, can be found in the complete, expanded version of this book, *Viability, Complexity and Us: The Framework Constraining Our Future* (Knight 2019). The introduction provides guidance about customizing its use.

ACKNOWLEDGEMENTS

All of you who indirectly contributed to this abridged version of *Viability, Complexity and Us: The Framework Constraining Our Future* are thanked in that book. I would not have completed this book without the help and support from Kathy and Alice, and the professionals who completed the production processes of editing, Emma Woodley, layout, Meghan Behse, and proofing, Toby Keymer.